The THz Dynamics of Liquids Probed by Inelastic X-Ray Scattering

Recommended Titles in Related Topics

X-Ray Scattering from Semiconductors and Other Materials
Third Edition
by Paul F Fewster
ISBN: 978-981-4436-92-2

Physics of Electrons in Solids
by Jean-Claude Tolédano
ISBN: 978-1-78634-972-9

Fundamentals of Quantum Materials
A Practical Guide to Synthesis and Exploration
edited by J Paglione, N P Butch and E E Rodriguez
ISBN: 978-981-121-936-8

The THz Dynamics of Liquids Probed by Inelastic X-Ray Scattering

Alessandro Cunsolo
University of Wisconsin-Madison, USA

World Scientific

NEW JERSEY · LONDON · SINGAPORE · BEIJING · SHANGHAI · HONG KONG · TAIPEI · CHENNAI · TOKYO

Published by

World Scientific Publishing Co. Pte. Ltd.

5 Toh Tuck Link, Singapore 596224

USA office: 27 Warren Street, Suite 401-402, Hackensack, NJ 07601

UK office: 57 Shelton Street, Covent Garden, London WC2H 9HE

Library of Congress Cataloging-in-Publication Data
Names: Cunsolo, Alessandro, author.
Title: The THz dynamics of liquids probed by inelastic X-ray scattering /
    Alessandro Cunsolo, University of Wisconsin-Madison, USA.
Description: New Jersey : World Scientific, [2020] | Includes bibliographical references and index.
Identifiers: LCCN 2020029887 | ISBN 9789813229488 (hardcover) |
    ISBN 9789813229495 (ebook) | ISBN 9789813229501 (ebook other)
Subjects: LCSH: X-rays--Scattering. | Terahertz spectroscopy. |
    Inelastic scattering. | Thermodynamics.
Classification: LCC QC482.S3 C86 2020 | DDC 532/.2--dc23
LC record available at https://lccn.loc.gov/2020029887

British Library Cataloguing-in-Publication Data
A catalogue record for this book is available from the British Library.

For any available supplementary material, please visit
https://www.worldscientific.com/worldscibooks/10.1142/10695#t=suppl

Typeset by Stallion Press
Email: enquiries@stallionpress.com

Printed in Singapore

Acknowledgements

In the preparation of this manuscript, I could count on the help of several people to whom I feel especially grateful. First, I would like to express my gratitude to my brotherly friend Alessio, who helped me in the critical revision of the book contents, engaging himself in the thorough proofreading of various book chapters. An invaluable help also came from Eleonora Guarini, Ubaldo Bafile and Nando Formisano who spotted several mistakes and weaknesses in the narrative, providing advice and guidance without parsimony. I'm grateful to Simone De Panfilis for the assistance in the preparation of the book cover and for sending me the picture appearing in it. A warm acknowledgement goes to the editor, Yong Qi Soh, for his precious support in this long-lasting endeavour. I'm especially indebted to Paola Vinciguerra, from whom I found the support, energy and motivation necessary to take essential steps of my life and work. My heartfelt gratitude is for my wife, Tawandra, and children, Matteo and Gabriel, for being present in my life, and for getting along with my frequent full immersions in the book editing. I also thank my brother, Antonio, and mother, Lucia Pratolini, for their continuous support from overseas. My first mentor, Michele Nardone, also deserves a special mention for having introduced me in this field of research. Finally, my deepest gratitude goes to my father, Salvatore Cunsolo, as he never gave up on the hope I could find in my work rewards at least comparable to those he had experienced as an accomplished physicist throughout his career.

Contents

Chapter 7:   The Q-evolution of the spectral shape from the hydrodynamic to the kinetic regime   169

Symbols used throughout the text

$A(t), \delta A(t)$	Generic dynamic variable and corresponding fluctuation
$\boldsymbol{A}(t), \delta \boldsymbol{A}(t)$	Set of dynamical variables and corresponding fluctuation
$\boldsymbol{A}(Q,t)$ (or $\boldsymbol{A}_Q(t)$)	Fourier Q-component of $\boldsymbol{A}(t)$
$\boldsymbol{A}(\boldsymbol{r})$	Vector potential at the point $\boldsymbol{r}$
$a_{k,\varepsilon}, a_{k,\varepsilon}^{\dagger}$	Annihilation and creation operators (with the indexes k, ε labelling wavevector and polarizatio states)
b_i	Neutron scattering length of the ith atomic species
c	Speed of light in vacuum
c_v, c_p	Constant volume and constant pressure specific heats
$c_s, c_s(Q)$	Adiabatic sound velocity and its finite-Q generalization
$c_T, c_T(Q)$	Isothermal sound velocity and its finite-Q generalization
$C_{AB}(t)$	Time correlation function of A and B
$\boldsymbol{C}(t)$	Time correlation matrix
$C_L(Q,\omega), C_T(Q,\omega)$	Longitudinal and transverse current spectra
D	Diffusion coefficient
$D_T, D_T(Q)$	Thermal diffusivity and its finite-Q generalization

E	Energy transfer ($= \hbar\omega$)
E_i, E_f	Energies of the photon (neutron) before and after the scattering event
$\widehat{\varepsilon}_i, \widehat{\varepsilon}_f$	Polarization versors of the incident and scattered photons
e	Electron charge
$|F_e\rangle, |I_e\rangle (|F_n\rangle, |I_n\rangle)$	Final and initial states of the electron (corresponding symbols for the nucleus)
$f(Q)$	Atomic form factor
$F(Q)$	Molecular form factor
$F(Q,t)$	Intermediate Scattering Function
$(F_s(Q,t), F_d(Q,t))$	(corresponding self and distinct parts)
$F, F(t)$	Force acting on a Hamiltonian system
$f(t), \boldsymbol{f}(t)$	Random force in the generalized Langevin equation for the single and multiple variables case
$G(\boldsymbol{r},t)$	Van Hove Function
$(G_s(\boldsymbol{r},t), G_d(\boldsymbol{r},t))$	(respective self and distinct parts)
$g(r)$	Pair distribution function
$H(t), H(\Gamma)$	Generic Hamiltonian (either in the time domain or in the phase-space one).
H_0, H_{int}, H'	Equilibrium, interaction and out-of-equilibrium Hamiltonians
$h, \hbar(= h/2\pi)$	Planck constant and reduced Planck constant
$\boldsymbol{J}(Q,t)$	Microscopic flow vector
J	Photon flux
$\boldsymbol{k}_i, \boldsymbol{k}_f$	Photon wavevector in its initial (before scattering) and final (after scattering) states
k_B	Boltzmann constant
$K(t), \boldsymbol{K}(t), K_n(t)$	Memory function, memory matrix and n-order memory function (within the continued fraction expansion)
$m_L(Q,t), \tilde{m}(Q,\omega)$	Second-order generalized memory function in the time and frequency domain
l, l_E	Mean free path and Enskog mean free path

m_e, m_n, M, M_{ST}	Electron, neutron, atomic (molecular) and Sachs–Teller mass
$n, n(\boldsymbol{r}, t), n_s(\boldsymbol{r}, t)$	Macroscopic (average) number density and its microscopic and self counterparts
$n(\omega)$	Bose population factor
$n(r)$	Number of particles within a distance r from a tagged particle in a liquid
n_s	Number of scattering units for unit volume
N	Number of atoms (particles) in a system
N_i, N_f	Numbers of incident and scattered photons
P	Pressure
$\boldsymbol{Q}, Q(=|\boldsymbol{Q}|)$	Exchanged wavevector and its amplitude
r_0	Classical electron radius ($r_0 = e^2/m_e c^2$ in cgs units)
$r_i(t)(\boldsymbol{R}_i(t))$	Position of the ith particle (atomic nucleus) at the time t
$r_i(t), p_i(t)$	Generalized phase space (position and momentum) coordinates of the ith particle of a many-particle system
$r_{ij}(t)$	Vector connecting the ith and the jth particles at the time t
$S(Q, \omega)$	Dynamic structure factor
$(S_I(Q, \omega), S_C(Q, \omega))$	(respective coherent and incoherent parts)
$S_A(\omega)$	Spectral density of A
$S(Q)$	Static structure factor
T	Temperature
T_{AW}	Period of the acoustic wave
T_r	"In cage" rattling period
T_0	Timescale characteristic of a measurement, or average time of the measurement
t_s	Sample thickness
V	Volume
$\boldsymbol{v}_i(t)$	Velocity of the ith particle at the time t
v_0	Thermal velocity
$V(r)$	Interaction potential

W_D	Darwin width
Z	Atomic number
z	Laplace transform variable (not to be confused with the Cartesian coordinate z)
$Z(Z')$	Partition function at equilibrium (its out of equilibrium counterpart)
$z_s, z_s(Q)$	Damping coefficient in the hydrodynamic spectrum and its finite-Q generalization

Greek letters

β	Inverse temperature in energy units $(=1/k_B T)$
$\gamma, \gamma(Q)$	Specific heats ratio and its finite-Q generalization
$\Gamma, \Gamma(Q), \Gamma_0, \Gamma_0$	Hydrodynamic acoustic damping, Damping of the DHO shape, damping matrix in the memory function formalism, and coefficient of the instantaneous term of the memory function
Γ, Γ_t	Generic point in the phase space and its time-dependent (time-propagated) counterpart
$\delta(x)$	Dirac's delta function in the variable x
ε_A	Parity of the variable A upon time-reversal ($\varepsilon_A = 1$ or -1)
η_s, η_B, η_L	Shear, bulk and longitudinal viscosity
$\eta_L(Q,t), \eta_L(Q,\omega)$	Generalized longitudinal viscosity in the time and frequency domain
2θ	Scattering angle
λ_{AW}	Wavelength of the acoustic wave
λ_B	De Broglie wavelength
μ	Absorption coefficient
$\rho(\Gamma)$	Equilibrium probability distribution (or density matrix, for quantum systems)

σ	Distance of minimum for the Lennard-Jones potential
σ_A, σ_C	Absorption and coherent X-Ray cross-sections
σ_{HS}	Hard-sphere diameter
$\sum_B$	Beam cross-sectional area
τ, τ_T, τ_C	Relaxation time (of the viscosity or the moduli), thermal relaxation time and compliance relaxation time
τ'	Time offset of the correlation function
τ_{cor}	Correlation time
τ_{coll}	Intercollision time
$\chi_{AB}(t)$	Response function of the system
$\tilde{\chi}_{AB}(Q, z = i\omega)$ $= \chi'_{AB}(\omega) + i\chi''_{AB}(\omega)$	Complex susceptibility
χ_T	Isothermal compressibility
ψ	Ratio between relaxation and collisional times
$\psi_A(t)$	Normalized time autocorrelation function of the variable A
ω	Exchanged (angular) frequency
$\omega_s, \omega_s(Q)$	Acoustic frequency and its finite Q generalization
Ω_0	Einstein frequency
Ω (or $\Delta\Omega$)	Solid angle
Ω	Frequency matrix of the generalized Langevin equation
$\Omega_c(Q)$	Frequency of the maximum in the current spectrum

Other symbols

$\frac{d\sigma}{d\Omega}$	Differential cross-section
$\frac{d^2\sigma}{d\Omega dE_f}$	Double differential cross-section
$\langle K.E. \rangle$	Mean single-particle kinetic energy
$\langle \Delta r^2(t) \rangle$	Mean square displacement
$f_B(\boldsymbol{k}_i, \boldsymbol{k}_f)$	Scattering length in Born approximation

$\mathcal{F}$	Free energy
$\mathfrak{L}$	Liouvillian operator
$\mathfrak{p}$	Mori-Zwanzig projection operator
$\mathfrak{q}$	Mori-Zwanzig orthogonal projection operator
∇, ∇^2	Nabla and Laplacian operators

Abbreviations

AI	Adiabatic-to-Isothermal transition
BLS, BUVS	Brillouin (visible) Light Ultra-Violet Scattering
CRZ	Compression-Rarefaction Zones (of an acoustic wave)
DAC	Diamond Anvil Cell
DHO	Damped Harmonic Oscillator
DINS, DIXS	Deep Inelastic Neutron and X-Ray Scattering
DLS	Depolarized Light Scattering
ESRF	European Synchrotron Radiation Facility
FWHM	Full Width at Half Maximum
GH	Generalized Hydrodynamics
HB	Hydrogen Bond
IA	Impulse Approximation
INS,IXS	Inelastic Neutron and X-Ray Scattering
KT	Kinetic Theory
LA, TA	Longitudinal and Transverse Acoustic (phonons)
MCT	Mode Coupling Theory
MD	Molecular Dynamics
MH	Molecular Hydrodynamics
PA	Poly(a)morphic (transition)
PSD	Positive Sound Dispersion
RTC	Rotational-Translational Coupling
US	Ultrasound Spectroscopy
VACF	Velocity Autocorrelation Function

Chapter 1

Introductory topics

The collective dynamics of disordered systems is a subject of prominent scientific interest since the dawn of modern science. However, despite an intensive theoretical, experimental and computational scrutiny, it still presents many puzzling aspects. The main reasons for this limited understanding are the lack of translational invariance in the microscopic structure of amorphous materials and the often exceptionally complex movements of their microscopic constituents.

From the experimental side, spectroscopic measurements can shed insight into this complicated issue. In this kind of measurements information on the internal movements of the sample can be achieved, for instance, by shooting it with a beam of particles[1] (photons, neutrons or electrons) of a given energy and then detecting the number of particles deviated by a given angle as a function of the energy they have gained or lost in the scattering process. From the energy conservation law, such energy equals the one either transferred to or acquired from the target sample. Therefore, the energy distribution (or the spectrum) of the scattered intensity conveys unique information on the various dynamical processes occurring inside the target sample. The key variables defining this distribution are the momentum, and the energy exchanged in the

---

[1]Notice that in some cases it might be convenient to vary the energy of the incident beam while keeping the scattering beam fixed, or performing angular scans while keeping fixed the energy difference between incident and scattering beam and so forth.

scattering event, i.e. $\hbar Q$ and $\hbar \omega$ respectively, with $\hbar$ being the reduced Planck constant. Usually, the dynamic variables Q and ω are respectively denominated the transferred wavevector and frequency.

In general, the measurement of the spectrum from a given sample, for instance, a liquid, enables us to probe dynamical events occurring inside such a sample as an average over a distance and timescale window limited to about $1/|Q|$ and $1/\omega$ respectively. Since both quantities increase upon increasing the wavelength of the incident quanta, short-wavelength radiation, such as X-Rays, enables us to probe length and timescales shorter by few orders of magnitudes than those accessible with visible light. This dynamic regime is customarily referred to as "mesoscopic" and corresponds to distances and times comparable with first neighbouring molecules separations and the periods of single molecules rattling within the cage formed by their first neighbours.

Conversely, visible light photons enable the access to the so-called "continuous" or "hydrodynamic" limit in which the fluid appears as a homogeneous and isotropic medium, and its transport behaviour is probed by averaging over a large number of microscopic events. On a general ground, a scattering process arises from the presence of inhomogeneities in the target sample; these can either have a static character, reflecting, for instance, the atomistic structure of matter, or a dynamic one, relating to the motions of atoms or molecules. Since Q and ω are the respective Fourier conjugate variables of space and time, static fluctuations primarily affect the Q dependence of the scattered radiation, while dynamic ones determine the ω dependence. However, in a strongly interacting system as a liquid, structural and dynamic properties are often entangled and their discrimination is sometime far from obvious.

In most cases, the sample can be thought of as a many-body system at thermal equilibrium. Although the fluid is at equilibrium over macroscopic scales, spontaneous fluctuations from such an equilibrium continuously occur and may have the form of propagating density waves, as well as of diffusive and relaxation processes.

Also, it is generally safe to assume that the photon beam induces a weak perturbation on the target sample, which in turn reacts linearly to such a disturbance. In these conditions, the linear response theory applies to the description of the scattering event. Although a formal derivation of the linear response theory goes beyond the scope of this book, it will be shown that this theory predicts that the dynamic response of the sample can be described in terms of its equilibrium properties, regardless of the nature of the applied perturbation. According to the so-called Onsager regression principle (Onsager, 1931a, Onsager, 1931b), the recovery of a system to the equilibrium after a weak macroscopic perturbation is governed by the same laws determining the relaxation of spontaneous microscopic fluctuations. Furthermore, within the validity of the fluctuation-dissipation theorem (Kubo, 1966), a spectroscopic measurement ultimately probes the dynamics of spontaneous fluctuations in a sample at equilibrium. As shown in Chapter 2, these density fluctuations can be investigated through the spectrum of their equilibrium correlation functions.

In fact, in the linear response approximation, the response of the system depends only parametrically on the nature of the induced perturbation. For instance, in ultrasound spectroscopy (US), mechanical perturbations are produced via piezoelectric transducers, in dielectric spectroscopy by the electric field between condenser plates, while in scattering experiments the disturbance arises from the probe-sample collision. Regardless of the nature of the weak perturbation applied, conventional spectroscopic measurements convey insight on an equilibrium property of the target sample, hereafter referred to as the sample's response.

References

Kubo, R. 1966. *Rep. Prog. Phys.*, **29**, 255.
Onsager, L. 1931a. *Phys. Rev.*, **37**, 405.
Onsager, L. 1931b. *Phys. Rev.*, **38**, 2265.

Chapter 2

Correlation functions and spectra

This chapter discusses some general properties of correlation functions of stochastic variables, which are observables directly accessed by scattering measurements. Correlation functions are mathematical entities, and, to some extent, their properties need to be discussed using a formal language. Nonetheless, these pages will give more emphasis to essential physical concepts, while the reader is often referred to the vast literature in the field for rigorous analytical derivations.

Spectroscopic methods treated in this book do not measure correlation functions directly in the space-time domain, but rather in the reciprocal space identified by the dynamic variables Q and ω introduced in the previous chapter. Therefore, the physical properties of relevance for spectroscopic experiments will be often represented here in such a reciprocal space.

It is essential to keep in mind that Q and ω are ideal variables to represent the outcome of spectroscopy results, as the response of a given sample to the "perturbation" induced by a scattering probe solely depend on them.

2.1. Spectroscopic experiments and correlation functions (Berne and Pecora, 1976)

Spectroscopic measurements discussed throughout this book probe fluctuations of some properties of a system at equilibrium. These fluctuations are erratic in both time and space, and they are probed by the spectroscopic measurement as averages over time lapses and distances much longer than those characterizing their rapid variations. To understand how this averaging works from a general perspective, it is useful to rest on some assumptions on the nature of the physical system under study.

A useful and relatively general assumption made throughout this book is that the system is ergodic. To introduce the concept of ergodicity, we may start from considering that the instantaneous state of an N-atoms system having $f \leq 3N$ degrees of freedom can be entirely specified by the f generalized coordinates $r_i(t)$ and momenta $p_i(t)$ of its microscopic constituents. At a given instant t^*, the state of the system is identified by a point in the phase space $(r_i(t^*), p_i(t^*))$. As time flows, the point moves, drawing trajectories which map energetically compatible (e.g. constant energy) surfaces of the phase space. In ergodic systems, these phase space surfaces are mapped uniformly, i.e. any arbitrarily small neighbourhood of these surfaces is crossed by an infinite number of trajectories. A prerogative of ergodic systems useful for the present scope is that averages of their time-dependent properties can be interchangeably performed either on the time or the phase space. The difference between an ergodic and a non-ergodic system is pictorially represented in the sketch of Figure 2-1.

After these preliminary remarks, let us now discuss a scenario of more direct pertinence for a spectroscopic experiment. Let us assume that a force field (or a perturbation) $F(t)$, acts on a sample originally at equilibrium and couples with a given sample property, $A(t)$. In a scattering experiment, it is natural to associate such a force field to the radiation beam impinging on the target sample.

Due to the existing coupling, a fluctuation of $A(t)$, hereafter referred to as $\delta A(t)$, causes a proportional fluctuation of $\delta F(t)$ and vice versa. Therefore:

$$\delta F(t) = K \delta A(t), \tag{2.1}$$

where K is a coupling constant.

As the system is initially at equilibrium, the property $A(t)$ does not depend on time explicitly but acquires its time dependence via the f pairs of coordinates and momenta of its microscopic constituents, i.e. $A(t) = A(r_i(t), p_i(t))$.

Non ergodic system   Ergodic system

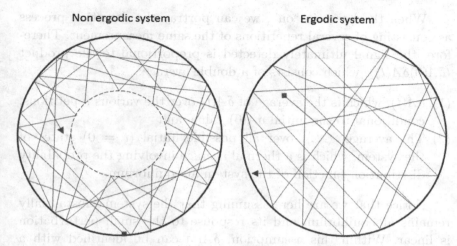

Figure 2-1: A schematic comparison between the phase-space trajectories representing the state of a non-ergodic (left) and an ergodic system (right). Notice that only the area schematically represented in the left is uniformly crossed by phase space trajectories.

Let us further assume, as appropriate in the specific context of this book, that the scattering measurement uses photons as probe particles; then the forcing term is to be identified with the electromagnetic field associated to the photon beam impinging on the sample.

The fluctuation of the force field $\delta F(t)$ results in the scattering of impinging photons. A detector, placed downstream to some energy analyzing device, ultimately assesses such a fluctuation, yet not by measuring $\delta F(t)$ directly, as it varies rapidly and erratically, but rather measuring a quantity proportional to its mean square value. A sensible question is how the latter relates to the fluctuations of the sample property $\delta A(t)$.

To answer this question, we will assume that the perturbing field is "switched on" at a time $t = 0$, at which the fluctuation of the system property $\delta A(0) = \delta A(r_i(0), p_i(0))$ is, in principle, well-defined, albeit generally unknown. Conversely, its time evolution $\delta A(t) = \delta A(r_i(t), p_i(t))$ is practically unpredictable, as it depends on the atomic coordinates and momenta of a system with many degrees of freedom.

When the beam is "on", we can portray the detection process as consisting of several repetitions of the same measurement. Therefore, the signal ultimately detected is proportional to the product $\langle \delta A(0) \delta \bar{A}(t) \rangle$, which consists of a double average:

(1) $\delta \bar{A}(t)$, which is the average of $\delta A(t)$ over the various repetitions, conditional to the initial $\delta A(0)$ value, and
(2) The average, $\langle \ldots \rangle$, over all possible initial $(t = 0)$ states of the system, which is a thermal average involving the probability distribution function of the system at equilibrium.

Notice that we are here assuming that the system substantially remains at equilibrium and its response to the small perturbation is linear. Within this assumption, $\delta A(t)$ can be identified with a spontaneous equilibrium fluctuations, as discussed further in depth in the next sections. If the considered system is ergodic, a direct route to evaluate the above average makes use of its stationary character, which stems from the circumstance that equilibrium distribution functions cannot depend on time. Hence, the measurement is proportional to the quantity:

$$R_{T_0}(t) = \frac{1}{T_0} \int_0^{T_0} dt_0 \delta A(t_0 + t) \delta A(t_0),$$

where T_0 is a timescale characteristic of the specific measurement. Notice that the average above, being an equilibrium property, should not depend on the initial time t_0 and, indeed, the integration runs over this variable. Within the assumption that the experiment performs an average over a time T_0 much longer than the timescale of single $\delta A(t)$ oscillations, the system response probed by the measurement becomes proportional to the infinitely long times limit of the above equation. Namely:

$$R_{T_0} = \langle \delta A(\tau') \delta A(0) \rangle = \lim_{T_0 \to \infty} \frac{1}{T_0} \int_0^{T_0} dt \delta A(t + \tau') \delta A(t), \quad (2.2)$$

where the variables in the integrand have been renamed for convenience.

The quantity defined above is the time autocorrelation function of the fluctuation $\delta A(t)$, and it only depends on the time offset τ'.

As we will see explicitly in the next chapter, the stochastic variable, A(t), of interest for the Inelastic X-Ray Scattering (*IXS*) technique is the microscopic density fluctuation, whose explicit expression is given at the end of this chapter.

To frame the concept of the correlation function in more intuitive terms, we can consider the irregular path traced by a fluctuating random variable and compare it with its copy rigidly shifted in time by an amount τ', as schematically illustrated in Figure 2-2. Although $A(t)$ undergoes rapid random fluctuations, if τ' is sufficiently small, all points belonging to the $A(t)$ and $A(t + \tau')$ profiles are likely to have similar values, i.e. the two curves are highly correlated. This is likely to be the case when τ' is much smaller than the timescale of any fluctuation of $A(t)$.

Conversely, if the time offset of the two profiles, τ', is much larger than the typical time of the $A(t)$ oscillations, no link is likely to exist between $A(t)$ and $A(t + \tau')$, i.e. the two profiles become statistically uncorrelated. It is thus reasonable to expect that the correlation between two identical fluctuating time profiles diminishes upon increasing their offset τ'.

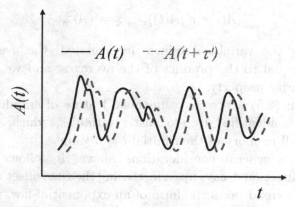

Figure 2-2: Schematic outline of the time evolution of a stochastic variable $A(t)$ and the corresponding time offset variable $A(t + \tau')$.

To describe this gradual loss of correlation quantitatively, one can multiply the values of the shifted pattern and that of the original dataset and compute the average of such a product over the time lapse, T_0. As mentioned in the previous chapter, the latter is a characteristic timescale of the experiment defining the extent of the probed time lapse. Hence:

$$\langle A(t + \tau')A(t) \rangle = \lim_{T_0 \to \infty} \frac{1}{T_0} \int_0^{T_0} dt A(t + \tau')A(t), \qquad (2.3)$$

where, again, the passage to the limit is justified by the fact that T_0 is much longer than the timescale of any $A(t)$ fluctuation. As noticed, the autocorrelation function depends uniquely on the time shift τ'. Specifically, if $\tau' = 0$, the two data sets are coincident and the average value,

$$\langle A(t + \tau')A(t) \rangle_{\tau'=0} = \langle A^2 \rangle, \qquad (2.4)$$

coincides with the mean square value of $A(t)$. In the formula above, we used the stationarity of equilibrium averages, which implies $\langle A^2(t) \rangle = \langle A^2 \rangle$; furthermore, we used the shorthand notation $A(0) = A$.

For $\tau \to \infty$, it is instead reasonable to expect that no correlation exists between the value that the variable assume at the time 0 and ∞, hence:

$$\langle A(t + \tau')A(t) \rangle_{\tau' \to \infty} = \langle A \rangle^2, \qquad (2.5)$$

since, being the variable's values uncorrelated, the average of a product is equal to the product of the averages, and we have used the stationarity property.

Equation (2.5) expresses a universal feature of ergodic systems: their state is identified by stochastic variables, which, after long times, lose all memory of their initial $t = 0$ value.

From the general considerations above, it follows that the correlation function decays upon increasing the time offset τ'. In some cases, such a trend takes the form of an exponential law, namely:

$$\langle A(\tau')A(0) \rangle = [\langle A^2 \rangle - \langle A \rangle^2] \exp(-\tau'/\tau_{cor}) + \langle A \rangle^2. \qquad (2.6)$$

The timescale τ_{cor} is referred to as the correlation time or, equivalently, the relaxation time of the variable $A(t)$.

As mentioned in the beginning of this paragraph, the crucial variable for a scattering measurement, rather than $A(t)$ itself, is its fluctuation from the equilibrium average, $\delta A(t) = A(t) - \langle A \rangle$. Assuming an exponential decay, the autocorrelation function of this new variable assumes the more compact form:

$$\langle \delta A(\tau')\delta A(0) \rangle = \langle (\delta A)^2 \rangle \exp(-\tau'/\tau_{cor}). \qquad (2.7)$$

This time autocorrelation function often appears in its normalized form $\psi_{\delta A}(\tau') = \langle \delta A(\tau')\delta A(0) \rangle / \langle (\delta A)^2 \rangle$, for which one has:

$$\psi_{\delta A}(\tau') = \exp(-\tau'/\tau_{cor}). \qquad (2.8)$$

Figure 2-3 illustrates a typical exponential time-decay of the correlation function.

The correlation between two distinct variables $A(t)$ and $B(t)$ is a generalization of the autocorrelation function, namely:

$$C_{AB}(\tau') = \langle A(t + \tau')B(t) \rangle = \langle A(\tau')B(0) \rangle, \qquad (2.9)$$

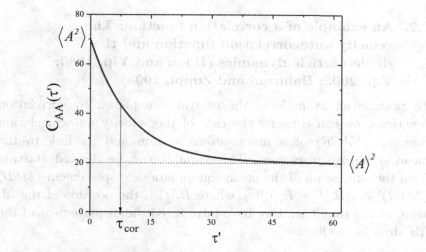

Figure 2-3: The time profile of an exponentially decaying time autocorrelation function (see text).

where the second equality reflects the stationarity of the equilibrium average.

Consistent with this notation, the autocorrelation function will be expressed as $C_{AA}(\tau')$.

The autocorrelation function is here extended to the case of complex variables as follows:

$$C_{AA}(\tau') = \langle A(\tau')A^*(0)\rangle,$$

where A^* is the complex conjugate of the complex variable A.

We notice that, when dealing with quantum systems, the time evolution of the autocorrelation function can be described using the Heisenberg representation:

$$\langle A(t)A^*(0)\rangle = Tr\left[\boldsymbol{\rho}(\boldsymbol{\Gamma})\exp\left(\frac{iH(\boldsymbol{\Gamma})t}{\hbar}\right)A(\boldsymbol{\Gamma})\right.$$

$$\left.\times\exp\left(-\frac{iH(\boldsymbol{\Gamma})t}{\hbar}\right)A^*(\boldsymbol{\Gamma})\right], \qquad (2.10)$$

where $\boldsymbol{\Gamma}$ is the point in the phase space describing the equilibrium state of the system, while $\boldsymbol{\rho}(\boldsymbol{\Gamma})$ is the corresponding density matrix.

2.2. An example of a correlation function: The velocity autocorrelation function and the single-particle dynamics (Boon and Yip, 1980; Yip, 2003; Balucani and Zoppi, 1994)

To provide an example of the central role played by correlation functions, we can consider the case of the velocity autocorrelation function (*VACF*) of a many atoms system and its link to the mean square displacement. This variable can be derived starting from the expression of the mean square atomic displacement (*MSD*) $\langle\Delta^2 r(t)\rangle = \langle[\boldsymbol{R}_i(t) - \boldsymbol{R}_i(0)]^2\rangle$, where $\boldsymbol{R}_i(t)$ is the position of the ith atom at the time t, and we have introduced the displacement of the ith atom in the fluid:

$$\boldsymbol{R}_i(t) - \boldsymbol{R}_i(0) = \int_0^t \boldsymbol{v}_i(t')dt'. \qquad (2.11)$$

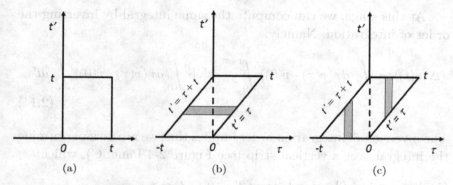

Figure 2-4: Schematic of the integration domains used to evaluate the mean square displacement integral (see text). Reproduced from Ref. (Yip, 2003).

Therefore, the mean square displacement reads as:

$$\langle \Delta^2 r(t) \rangle = \int_0^t dt' \int_0^t \langle v(t') \cdot v(t'') \rangle dt''. \qquad (2.12)$$

Notice that the integration domain of the integral in Eq. (2.12) is a square (see Figure 2-4, Panel A) of size t. The term under integral is the velocity autocorrelation function. Since a fluid at equilibrium is indeed a homogeneous system, one can use the stationarity of the correlation function, which yields:

$$\langle v(t') \cdot v(t'') \rangle = \langle v(t' - t'') \cdot v(0) \rangle.$$

At this stage, we can change the variable of integration to $\tau = t' - t''$, thus obtaining:

$$\langle \Delta^2 r(t) \rangle = \int_0^t dt' \int_{t'-t}^{t'} \langle v(\tau) \cdot v(0) \rangle d\tau. \qquad (2.13)$$

With this change of variables, the integration domain is transformed into a parallelogram, which is "sliced" along an infinitesimal horizontal stripe, extending between the parallel lines $\tau = t' - t$ and $\tau = t'$ (rightmost integral); in the integration, such a stripe moves between 0 and t. (see Figure 2-4 Panel B).

At this stage, we can compute the same integral by inverting the order of integration. Namely:

$$\langle \Delta^2 r(t) \rangle = \int_{-t}^{0} d\tau \langle \boldsymbol{v}(\tau) \cdot \boldsymbol{v}(0) \rangle \int_{0}^{t+\tau} dt' + \int_{0}^{t} d\tau \langle \boldsymbol{v}(\tau) \cdot \boldsymbol{v}(0) \rangle \int_{\tau}^{t} dt'.$$
(2.14)

The inversion of integration order implies that we are now performing the integral over a vertical strip (see Figure 2-4 Panel C), which:

(1) For $t \in [-t, 0]$ extends from $t' = 0$ to $t' = t + \tau$;
(2) For $t \in [0, t]$ extends from $t' = \tau$ to $t' = t$.

In the right hand side of Eq. (2.14), the two rightmost integrals in the variable t' can be further developed to obtain:

$$\langle \Delta^2 r(t) \rangle = \int_{-t}^{0} d\tau \langle \boldsymbol{v}(\tau) \cdot \boldsymbol{v}(0) \rangle (t + \tau) + \int_{0}^{t} d\tau \langle \boldsymbol{v}(\tau) \cdot \boldsymbol{v}(0) \rangle (t - \tau).$$
(2.15)

In the first integral on the right-hand side, the integration variable can be changed from τ to $-\tau$, while considering that $\langle \boldsymbol{v}(0) \cdot \boldsymbol{v}(\tau) \rangle = \langle \boldsymbol{v}(0) \cdot \boldsymbol{v}(-\tau) \rangle$, due to the time evenness of classical autocorrelation functions (see footnote 1 further below). Therefore, the two integrals are equal, and one has:

$$\langle \Delta^2 r(t) \rangle = 2 \int_{0}^{t} d\tau (t - \tau) \langle \boldsymbol{v}(\tau) \cdot \boldsymbol{v}(0) \rangle.$$
(2.16)

The above formula can be written in the form:

$$\langle \Delta^2 r(t) \rangle = 6 v_0^2 \int_{0}^{t} (t - \tau) \psi_v(\tau) d\tau,$$
(2.17)

where $v_0 = \sqrt{k_B T / M}$ is the thermal velocity, M the atomic mass, and $\psi_v(t) = \langle \boldsymbol{v}(\tau) \cdot \boldsymbol{v}(0) \rangle / \langle \boldsymbol{v}(0) \cdot \boldsymbol{v}(0) \rangle$ the normalized *VACF*.

For long times, one has:

$$\langle \Delta^2 r(t) \rangle = 6 D t,$$
(2.18)

where:

$$D = v_0^2 \int_0^\infty \psi_v(\tau) d\tau \tag{2.19}$$

is the diffusion coefficient.

The above equation is an example of the so-called Green–Kubo formulas, which set a link between transport coefficients and correlation functions at equilibrium.

For short times, the *VACF* in Eq. (2.19) can be expressed as a time expansion of the form:

$$\langle \boldsymbol{v}(t) \cdot \boldsymbol{v}(0) \rangle = \frac{3k_B T}{M} - \langle |\dot{\boldsymbol{v}}(0)|^2 \rangle \frac{t^2}{2} + \cdots , \tag{2.20}$$

where we exploited the time evenness of classical autocorrelation functions[1] and considered that, from the equipartition theorem,

---

[1] We can demonstrate this properties in two steps, while considering the definition of correlation function in the classical case.

(1) Using the invariance upon time shift one has that:

$$C_{AA}(t) = \varepsilon_A C_{AA}(-t) = \varepsilon_A \langle A(-t) A^*(0) \rangle = \varepsilon_A \langle A(0) A^*(t) \rangle = \varepsilon_A C_{AA}^*(t),$$

where ε_A (either $= 1$ or -1) defines the time-reversal parity of $A(t)$ i.e. $A(-t) = \varepsilon_A A(t)$.

The comparison of the first and the last member of the identity chain implies that if $\varepsilon_A = -1$, then $C_{AA}(t) \equiv 0$, while if $\varepsilon_A = 1$ then $C_{AA}(t) \equiv C_{AA}^*(t)$, i.e. $C_{AA}(t)$ is real.

(2) By definition:

$$C_{AA}(\tau') = \lim_{T \to \infty} 1/T \int_0^T A(t + \tau') A^*(t) dt$$

$$= \lim_{T \to \infty} 1/T \int_0^T A(t) A^*(t - \tau') dt = C_{AA}^*(-\tau') = C_{AA}(-\tau')$$

where the second identity follows from the invariance upon time shift (where the time shift is $-\tau'$) and the last from the fact that $C_{AA}(\tau)$ is real, as demonstrated in (1). In summary, $C_{AA}(t) = C_{AA}(-t)$.

Combining the points (1) and (2) we have demonstrated that the autocorrelation function (of real variables) is a real and even function of time. This demonstration does not apply to the quantum autocorrelation function (see Eq. 2.10), which is a complex number. Quantum correlation function cannot be directly measured, but observables are often related to their real or imaginary part or to suitable combinations of correlation functions.

$\langle |v(0)|^2 \rangle = 3k_B T / M$. Correspondingly, the mean square displacement defined through Eq. (2.17) can be expressed by the following short-time expansion of the *MSD*:

$$\langle \Delta^2 r(t) \rangle = \frac{3k_B T}{M} t^2 - \frac{k_B T}{4M} \Omega_0^2 t^4 + \ldots, \qquad (2.21)$$

whose explicit derivation is omitted here and the interested reader can refer to Ref. (Balucani and Zoppi, 1994). In the expansion above, Ω_0 is the so-called Einstein frequency of a liquid, defined by:

$$\Omega_0^2 = \frac{n}{3M} \int d\boldsymbol{r} \nabla^2 V(r) g(r), \qquad (2.22)$$

where n is the number density of the system, while $V(r)$ is the pair potential acting between two atoms at a distance $r = |\boldsymbol{r}|$ with $\boldsymbol{r}$ being the vector connecting the centres of the two atoms. Indeed, unless otherwise stated, it will be generally assumed throughout this book that all microscopic interactions in the fluid are pairwise and additive. In the above equation, we have introduced the pair correlation function $g(r)$ — to be further discussed at the end of this chapter — which represents the distribution probability to have an atom at a distance r from a tagged atom located at the origin. Furthermore, Eq. (2.22) contains the Laplacian operator defined as the sum of the second unmixed derivatives with respect to the Cartesian components, that is $\nabla^2 = \nabla \cdot \nabla = \frac{\partial^2}{\partial x^2} + \frac{\partial^2}{\partial y^2} + \frac{\partial^2}{\partial z^2}$. Notice that owing to the isotropic nature of the fluid system, $V(r)$ and $g(r)$ do not depend on the direction of $\boldsymbol{r}$, but uniquely on its amplitude, r.

Equation (2.21) indicates that, for short times, the *MSD* is proportional to t^2, as predicted by the equation of motion for a free, non-interacting particle. After a time comparable with the inverse of the Einstein frequency, a deviation of the *MSD* from its t^2-behaviour is observed. For a system resembling an hard sphere gas, i.e. a system whose dynamics is primarily governed by mutual collisions between rigid particles, Ω_0^{-1} yields a measure of the time elapsed between two successive collisions. In general, for real fluids, the variable Ω_0 is expectedly simply connected to the average rattling frequency of the single atom inside the first neighbours' cage.

In summary, the different regimes of the single particle motion are schematically represented in Figure 2-5. For short times, the dynamic

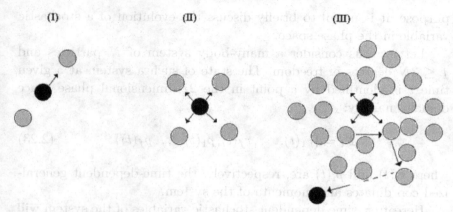

(I)   (II)   (III)

Figure 2-5: Schematic representation of the single-particle motion at three different time scales: (I) Single-particle or ballistic; (II) Cage oscillation and (III) Diffusive regimes.

event which primarily influences the MSD is the free streaming of the tagged particle between successive collisions within the first shell of neighbours. At longer times the atomic movement is dominated by "in cage" rattling movements, with frequency $\sim \Omega_0$.

For times exceeding Ω_0^{-1} the first neighbour's cage surrounding the tagged atom relaxes, making allowance for its diffusive "out-of-the-cage" jumps.

Finally, in the long-time limit, the particle has experienced a significant number of collisions, and the transition to the diffusive regime is characterized by a linear t dependence of the MSD.

In the intermediate range corresponding to times comparable with the inverse of the Einstein frequency, the single-particle motion has a vague resemblance with that of a molecule in a crystal lattice (Balucani and Zoppi, 1994), as it mostly consists of oscillations around the cage centre, with a frequency determined by the second derivative of the potential.

2.3. Time averages and ensemble averages (Berne and Pecora, 1976)

As mentioned in section 2.1, for ergodic systems, it might be convenient to compute correlation functions as phase space, rather than time averages, given that the two averages equivalent. To this purpose, it is useful to briefly discuss the evolution of a stochastic variable in the phase space.

Let us thus consider a many-body system of N particles and $f \leq 3N$ degrees of freedom. The state of such a system at a given time t is identified by a point in the $2f$-dimensional phase space domain, namely:

$$\boldsymbol{\Gamma}_t = \{r_1(t), \ldots, r_f(t), p_1(t), \ldots, p_f(t)\}, \qquad (2.23)$$

where $r_i(t)$ and $p_i(t)$ are, respectively, the time-dependent generalized coordinates and momenta of the system.

Hereafter, time-dependent stochastic variables of the system will be often labelled using the implicit notation $A(t) \equiv A(\boldsymbol{\Gamma}_t)$. Let us now consider one of such variables; its time derivative reads as:

$$\frac{dA(\boldsymbol{\Gamma}_t)}{dt} = \frac{\partial A(\boldsymbol{\Gamma}_t)}{\partial t} + \sum_{i=1}^{N} \left[\frac{\partial A(\boldsymbol{\Gamma}_t)}{\partial r_i} \dot{r}_i + \frac{\partial A(\boldsymbol{\Gamma}_t)}{\partial p_i} \dot{p}_i \right] \qquad (2.24)$$

At this stage, when the system is at the equilibrium, we can assume that the variable A is stationary, i.e. the term $\partial A(\boldsymbol{\Gamma}_t)/\partial t = 0$. The partial derivatives in the above equation obey the so-called canonical equations:

$$\dot{r}_i = \frac{\partial H}{\partial p_i}$$
$$\qquad\qquad i = 1, \ldots, f \qquad (2.25)$$
$$\dot{p}_i = -\frac{\partial H}{\partial r_i}$$

where, to ease the notation, we omitted the reference to the time-dependence of the phase space coordinates. Also, here H is the Hamiltonian, of the system, which can be written as $H = \mathbb{T} + \mathbb{W}$; here $\mathbb{T} \equiv \mathbb{T}[p_1(t), \ldots p_f(t)]$ and $\mathbb{W} \equiv \mathbb{W}[r_1(t), \ldots r_f(t)]$ are the kinetic and the potential energies, respectively.

Once the initial state $\boldsymbol{\Gamma} = \{r_1(0), \ldots r_f(0), p_1(0), \ldots p_f(0)\}$ is specified, the canonical equations can be solved and the state of the system uniquely identified at any time.

Upon inserting Eq. (2.25) into Eq. (5.16), one obtains the following equation:

$$\frac{dA(\boldsymbol{\Gamma}_t)}{dt} = \sum_{i=1}^{f}\left[\frac{\partial H}{\partial p_i}\frac{\partial}{\partial r_i} - \frac{\partial H}{\partial r_i}\frac{\partial}{\partial p_i}\right]A(\boldsymbol{\Gamma}_t) = i\mathcal{L}A(\boldsymbol{\Gamma}_t), \qquad (2.26)$$

which defines the Liouville operator, or Liouvillian, $\mathcal{L}$.

In the time domain, the solution of the above equation reads as:

$$A(t) = \exp(i\mathcal{L}t)A, \qquad (2.27)$$

where, as usual, $A = A(0)$. The action of the time propagator, $\exp(i\mathcal{L}t)$, may appear more evident if one considers its series expansion $\exp(i\mathcal{L}t) = \sum_{n=1}^{\infty}(it)^n\mathcal{L}^n/n!$, where, for a given integer n, one has $\mathcal{L}^nA = \mathcal{L}[\mathcal{L}^{n-1}A]$; for instance, $\mathcal{L}^2A = \mathcal{L}[\mathcal{L}A]$, or $\mathcal{L}^3A = \mathcal{L}[\mathcal{L}[\mathcal{L}A]]$ and so forth. Equation (2.27) authorizes the expression of the correlation function between two variables A and B in terms of such a time propagator:

$$\langle A(t)B^*\rangle = \int d\boldsymbol{\Gamma}\rho(\boldsymbol{\Gamma})[\exp(i\mathcal{L}t)A]B^*,$$

where $\rho(\boldsymbol{\Gamma})$ is the equilibrium probability distribution function and $\rho(\boldsymbol{\Gamma})d\boldsymbol{\Gamma}$ the probability of finding the system in the initial state $\boldsymbol{\Gamma}$ to within a phase-space volume $d\boldsymbol{\Gamma}$. Of course, the specific form of $\rho(\boldsymbol{\Gamma})$ depends on the ensemble representative of the target system.

In Chapter 5, we will show how the Liouville equation can be handled analytically in the framework of the so-called Mori–Zwanzig formalism.

We now focus on formal aspects associated with correlation functions, which are of relevance for spectroscopy applications.

2.4. The Onsager's regression principle (Onsager, 1931a, Onsager, 1931, Batista)

The central role played by density fluctuations in scattering experiments can be better emphasized by introducing the Onsager regression principle. This principle states that if a weak macroscopic perturbation drives a system's property A away from its equilibrium value, its irreversible relaxation back to equilibrium is simply described by the (equilibrium) autocorrelation function of spontaneous microscopic fluctuations δA.

To formally express this principle, we can start from assuming that a weak macroscopic perturbation field F is "switched on" at $t = -\infty$ and this field couples with some microscopic property of the system $A(t)$; after an infinitely long time lapse (i.e. by the time $t = 0$), F has driven such a variable to the non-vanishing, off-equilibrium, value $A = A(0)$. In this $t = 0$ instant, the state of the system is described by the perturbed Hamiltonian $H' = H_{eq} + \Delta H$, which, due to the weakness of the perturbation, differs from its equilibrium counterpart, H_{eq}, only slightly. The off-equilibrium part ΔH, which arises from the coupling between F and A, can be expressed as a power expansion in the small variable F, which at the first order reads as:

$$\Delta H = -FA. \qquad (2.28)$$

As mentioned, at $t = 0$ the force F is "switched off" and $A(t)$ is allowed to relax back to its thermal equilibrium value irreversibly.

In the following, we will treat this relaxation process using the tools of classical statistical mechanics, although the treatment can be easily generalized to the quantum case.

We here assume that the evolution of the system is monitored through the variation of the same variable A the force couples with. For $t = 0$, the initial macrostate of the system can be described by introducing the macroscopic average:

$$\bar{A}(0) = \frac{\int d\boldsymbol{\Gamma} A(\boldsymbol{\Gamma}) \exp[-\beta H'(\boldsymbol{\Gamma})]}{Z'}, \qquad (2.29)$$

where we have introduced the non-equilibrium partition function $Z' = \int d\boldsymbol{\Gamma} \exp[-\beta H'(\boldsymbol{\Gamma})]$. On a rigorous ground, at $t = 0$ the system, rather than being out-of-equilibrium, is in equilibrium with the external force F and its state is described by the Hamiltonian H'.

For $t > 0$, i.e. after the force is removed, the relaxation of $A(t)$ back to its thermal equilibrium value can be described by introducing the following time-dependent non-equilibrium average:

$$\bar{A}(t) = \frac{\int d\Gamma A(\boldsymbol{\Gamma}_t) \exp[-\beta H'(\boldsymbol{\Gamma})]}{Z'}. \tag{2.30}$$

Hereafter throughout these paragraphs, the overbar and the angle brackets will be used to label averages in and out thermal equilibrium, respectively.

As demonstrated further below, the relaxation of $\bar{A}(t)$ assumes the form:

$$\frac{\bar{A}(t) - \langle A \rangle}{\bar{A}(0) - \langle A \rangle} = \frac{\langle \delta A(t) \delta A \rangle}{\langle (\delta A)^2 \rangle}, \tag{2.31}$$

where $\delta A(t)$ is the spontaneous microscopic fluctuation of the variable $A(t)$. The above equation — strictly valid when the system remains close to equilibrium, i.e. $\bar{A}(0) - \langle A \rangle$ is small — expresses quantitatively the Onsager's regression hypothesis, which links an externally induced macroscopic fluctuation of the system $(\bar{A}(t) - \langle A \rangle)$ to the correlation function of spontaneous equilibrium fluctuations $\delta A(t)$. To emphasize the pertinence of this principle to the case of spectroscopic experiments, one might simply relate F to the electromagnetic field carried by the photon beam. The variable $A(t)$ is to be identified with the variable coupled to such an electromagnetic field, i.e. ultimately the atomic (microscopic) density fluctuation. In fact, although the electromagnetic field directly couples only with the electronic clouds of the target system's atoms, within the Born–Oppenheimer approximation, these follows the slowly moving nuclear positions $\boldsymbol{R}_i(t)$ which determine the microscopic density fluctuations (see section 2.9).

A demonstration of the Onsager's principle (Batista)

It can be readily noticed that $\bar{A}(t)$ in Eq. (2.30) is obtained from $\bar{A}(0)$ in Eq. (2.29) by merely replacing $A(\boldsymbol{\Gamma})$ with $A(\boldsymbol{\Gamma}_t)$ in the integrand. For a small ΔH, the exponential factor determining the partition function can be written as:

$$\exp(-\beta H'(\boldsymbol{\Gamma})) = \exp[-\beta(H_{eq}(\boldsymbol{\Gamma}) + \Delta H(\boldsymbol{\Gamma}))]$$
$$\approx [1 - \beta \Delta H(\Gamma)] \exp[-\beta H_{eq}(\boldsymbol{\Gamma})]$$

Hence, the out-of-equilibrium average reads as:

$$\bar{A}(t) \approx \frac{\int d\boldsymbol{\Gamma}[1 - \beta \Delta H(\boldsymbol{\Gamma})] A(\boldsymbol{\Gamma}_t) \exp[-\beta H_{eq}(\boldsymbol{\Gamma})]}{\int d\boldsymbol{\Gamma}[1 - \beta \Delta H(\boldsymbol{\Gamma})] \exp[-\beta H_{eq}(\boldsymbol{\Gamma})]}$$
$$= \frac{\int d\boldsymbol{\Gamma}[1 - \beta \Delta H(\boldsymbol{\Gamma})] A(\boldsymbol{\Gamma}_t) \exp[-\beta H_{eq}(\boldsymbol{\Gamma})]}{Z - \beta \int d\boldsymbol{\Gamma} \Delta H(\boldsymbol{\Gamma}) \exp[-\beta H_{eq}(\boldsymbol{\Gamma})]},$$

where $Z = \int d\boldsymbol{\Gamma} \exp[-\beta H_{eq}(\boldsymbol{\Gamma})]$ is the equilibrium partition function. Therefore:

$$\bar{A}(t) \approx \frac{\int d\Gamma[A(\boldsymbol{\Gamma}_t) - \beta A(\boldsymbol{\Gamma}_t)\Delta H(\boldsymbol{\Gamma})] \exp[-\beta H_{eq}(\boldsymbol{\Gamma})]}{Z}$$
$$\times \frac{1}{1 - \frac{[\beta \int d\boldsymbol{\Gamma} \Delta H(\boldsymbol{\Gamma}) \exp[-\beta H_{eq}(\boldsymbol{\Gamma})]]}{Z}}.$$

By recognizing that the equilibrium average of a generic property x is $\langle x \rangle = \int d\boldsymbol{\Gamma} x(\boldsymbol{\Gamma}) \exp[-\beta H_{eq}(\boldsymbol{\Gamma})]/Z$, one can write:

$$\bar{A}(t) \approx [\langle A(t) \rangle - \beta \langle \Delta H(\boldsymbol{\Gamma}) A(t) \rangle] \frac{1}{1 - \beta \langle \Delta H(\boldsymbol{\Gamma}) \rangle}$$
$$\approx \langle A(t) \rangle (1 + \beta \langle \Delta H(\boldsymbol{\Gamma}) \rangle)$$
$$- \beta \langle \Delta H(\boldsymbol{\Gamma}) A(t) \rangle (1 + \beta \langle \Delta H(\boldsymbol{\Gamma}) \rangle),$$

where the second identity is derived using a first-order truncated expansion of the form $(1 - x)^{-1} \approx 1 + x$, valid for a generic small variable x. After some trivial manipulations and considering that $\Delta H = -FA$, one obtains:

$$\bar{A}(t) - \langle A \rangle \approx \beta F(\langle A(t)A \rangle - \langle A \rangle^2) \qquad (2.32)$$

and

$$\bar{A}(0) - \langle A \rangle \approx \beta F(\langle A^2 \rangle - \langle A \rangle^2), \qquad (2.33)$$

By dividing the last two equations member-by-member and considering that $\delta A(t) = \langle A(t)A \rangle - \langle A \rangle^2$, one obtains:

$$\frac{\bar{A}(t) - \langle A \rangle}{\bar{A}(0) - \langle A \rangle} = \frac{\langle \delta A(t) \delta A \rangle}{\langle (\delta A)^2 \rangle}, \qquad (2.34)$$

which coincides with the Onsager principle, as expressed by Eq. (2.31).

In summary, introducing the macroscopic fluctuations of the system $\Delta \bar{A}(t) = \bar{A}(t) - \langle A \rangle$, Eq. (2.32) can be rewritten as:

$$\Delta \bar{A}(t) = \beta F \langle \delta A(t) \delta A \rangle. \qquad (2.35)$$

The above formula shows that, once the constraint (i.e. the perturbation field) is removed, the relaxation of the macroscopic perturbation $\Delta \bar{A}(t)$ is proportional to the equilibrium time auto-correlation function of the spontaneous fluctuations $\delta A(t)$.

Hence, in the limit of a small macroscopic perturbation, also referred to as linear response regime, the knowledge of equilibrium properties, like correlation functions, is useful for predicting non-equilibrium transient processes.

The one above represents a particular case of the so-called Fluctuation-Dissipation theorem (Callen and Welton, 1951). In this particular case, the perturbation couples directly with the relaxing macroscopic variable $A(t)$. A more general case, in which the force couples with a generic variable, $B(t)$, will be considered in the next section.

2.5. The linear response of the system (MacKintosh)

In the previous paragraph, we have considered the response of the system to a constant force or perturbation field. The latter is switched on infinitely far in the past and by the time $t = 0$ it brings the system in an off-equilibrium state. Being this constant perturbation switched off at $t = 0$, it has the following stepwise time-dependence:

$$F(t) = \begin{cases} F & \text{for } t < 0 \\ 0 & \text{otherwise.} \end{cases} \qquad (2.36)$$

Let us now consider the more general case, in which the perturbing field coupling with A has an arbitrary time-dependence, non-necessarily coincident with the step down trend described above. Again, provided the interaction is weak, we can express the non-equilibrium average $\bar{A}(t)$ as a power series of $F(t)$, which, to the first order, reads:

$$\bar{A}(t) = \langle A \rangle + \int dt' \chi(t, t') F(t'), \qquad (2.37)$$

which defines the response function $\chi(t, t')$. Notice that the leading term in the expansion, i.e. the 0th order in $F(t)$, describes a situation in which the force is not applied, and the system is in its thermal equilibrium, where $\bar{A}(t) = \langle A \rangle$.

The expansion above is carried out assuming that the force acts at the time t' and one observes its effect at the time t, where integration expresses the dependence of $\bar{A}(t)$ on the full history of the interaction, which is embodied in the response function $\chi(t, t')$.

Again, if one considers the time-dependent fluctuation $\Delta \bar{A}(t) = \bar{A}(t) - \langle A \rangle$, the above equation transforms to:

$$\Delta \bar{A}(t) = \int dt' \chi(t, t') F(t'). \qquad (2.38)$$

In general the response $\chi(t, t')$ must fulfill a few physical requirements:

(1) Causality

The effect of the force can be perceived even from arbitrarily distant times, which implies that the response should be integrated over all times $t < 0$. However, since the response can only happen after the perturbation (causality principle), the integration domain cannot extend to times successive to the observation time $(t' > t)$. This implies that $\chi(t, t') = 0$ for $t' > t$. Hence, the equation above transforms to:

$$\Delta \bar{A}(t) = \int_{-\infty}^{t} dt' \chi(t, t') F(t'), \qquad (2.39)$$

The causality principle can be superimposed by writing the response as $\Theta(t - t')\chi(t, t')$, where the Heaviside step function $\Theta(t - t')$ is equal to 1 for $t \geq t'$ and 0 otherwise.

(2) Stationarity

The response is a property of the (equilibrium) system, which must not depend on the weak perturbing force. As any other equilibrium property, the response must be invariant for time shifts or, equivalently, it must uniquely depend on the time lapse between the application of the force and the time of observation, i.e. $\chi(t, t') = \chi(t - t')$. As a consequence, one has:

$$\Delta \bar{A}(t) = \int_{-\infty}^{t} dt' \chi(t - t') F(t'). \tag{2.40}$$

One can define the new variable $t - t' = \tau$, which enables one to write the above equation as:

$$\Delta \bar{A}(t) = \int_{0}^{\infty} d\tau \chi(\tau) F(t - \tau). \tag{2.41}$$

(3) Response to an impulse

Let us now consider the case of a force having an impulsive character, thus being consistently approximated by a δ-function of time $F(t) = F_0 \delta(t - t')$. Under this condition, one has:

$$\Delta \bar{A}(t) = F_0 \chi(t - t'),$$

which shows that χ describes the time disturbance induced on the system when an impulsive perturbation is applied. More in general, response functions describe how systems evolve after being perturbed by an impulsive force, i.e. a function representable as a delta-function of time.

If instead the force has the stepwise form of Eq. (2.36), i.e. if it is equal to F for negative times and vanishing elsewhere, Eq. (2.41) can be rewritten as:

$$\Delta \bar{A}(t) = F \int_{t}^{\infty} d\tau \chi(\tau),$$

which combined with Eq. (2.35), gives the following response function for $t > 0$:

$$\chi(t) \equiv \chi_{AA}(t) = -\beta\theta(t)\frac{d}{dt}\langle\delta A(t)\delta A(0)\rangle,$$

where, again, the presence of Heaviside step function $\theta(t)$ stems from the causality principle.

In general, provided the perturbing field (or the generalized force) F is weak, one can perform a series expansion of the Hamiltonian in power of such a field; the leading (linear) order of such an expansion reads as:

$$H' = H_{eq} - FB, \tag{2.42}$$

where the term linear in the field now depends on the stochastic variable B, i.e. we are here assuming that the variable the perturbation couple with is not necessarily equal to A. In this scenario, F and B are thermodynamically conjugate variables — as, for instance, pressure and volume or entropy and temperature — i.e. $F = -\partial F/\partial B$, where $\mathcal{F} = k_B T \ln Z$ is the free energy.

Again, the response of the system to the perturbation is monitored through the variation of the variable A, whose out-of-equilibrium average reads as:

$$\bar{A}(t) = \frac{1}{Z'}\int d\boldsymbol{\Gamma}\exp[-\beta H'(\boldsymbol{\Gamma})]A(\boldsymbol{\Gamma}_t), \tag{2.43}$$

with $Z' = \int d\boldsymbol{\Gamma}\exp[-\beta H'(\boldsymbol{\Gamma})]$, and $H'(\boldsymbol{\Gamma}) = H_{eq}(\boldsymbol{\Gamma}) - FB(\boldsymbol{\Gamma})$.

By taking the derivative of Eq. (2.43) with respect to the force one has:

$$\frac{d\bar{A}(t)}{dF} = \beta\left[\begin{array}{c}\dfrac{\int d\boldsymbol{\Gamma}\, A(\boldsymbol{\Gamma}_t)B(\boldsymbol{\Gamma})\exp[-\beta H'(\boldsymbol{\Gamma})]}{Z'} \\[1.5em] -\dfrac{\int d\boldsymbol{\Gamma}\, A(\boldsymbol{\Gamma}_t)\exp[-\beta H'(\boldsymbol{\Gamma})]}{Z'^2}\beta\int d\boldsymbol{\Gamma}\exp[-\beta H'(\boldsymbol{\Gamma})]B(\boldsymbol{\Gamma})\end{array}\right].$$

In particular, if such a derivative is computed in $F = 0$, where $H'(\boldsymbol{\Gamma}) = H_{eq}(\boldsymbol{\Gamma})$, the equation above transforms to:

$$\frac{d\bar{A}(t)}{dF}\bigg|_{F=0} = \beta[\langle A(t)B(0)\rangle - \langle A(0)\rangle\langle B(0)\rangle]$$

$$= \beta\langle \delta A(t)B(0)\rangle,$$

as deduced after recognizing that $\boldsymbol{\Gamma}$ and $\boldsymbol{\Gamma}_t$ are the time space points defining sample microstates at the times 0 and t respectively. Notice that, being the above expression obtained for $F = 0$, all averages and correlations appearing therein are computed at equilibrium and involve time-dependent quantities, i.e. spontaneous equilibrium fluctuations. Indeed, although the system is macroscopically out of equilibrium the $\boldsymbol{\Gamma} \to \boldsymbol{\Gamma}_t$ evolution of system's microstates is ruled by a time-propagator, $TP\{\dots\}$, which only depends on the equilibrium Hamiltonian $H_{eq}(\boldsymbol{\Gamma})$.

By integrating both members of the above equation, one eventually obtains:

$$\Delta\bar{A}(t) = -\beta F\langle \delta A(t)B(0)\rangle. \tag{2.44}$$

The one above is a more general formulation of the Fluctuation-Dissipation theorem than the one expressed by Eq. (2.35). Generalizing the previous treatment of the $A = B$ case, the response of the system for $t > 0$ can be expressed as:

$$\chi_{AB}(t) = -\beta\theta(t)\frac{d}{dt}\langle \delta A(t)B(0)\rangle. \tag{2.45}$$

Let us now go back to the case in which the force couples with the fluctuating variable A, for which the response can be expressed as in Eq. (2.42). Indeed, as discussed further below, this is a case of direct pertinence with a scattering measurement. Notice that, due to the discussed causality property, the left-hand side of Eq. (2.42) must vanish for $t < 0$, while, conversely, the correlation function is an even function of time. This imply that the given definition of the response function cannot be valid for negative times. This for instance causes a problem while attempting to represent, via Fourier transform, the response function in the frequency domain. It is useful to rewrite the definition of the response function in a way that makes it valid also for $t > 0$. To deal with response functions defined in all times, it is useful to write:

$$\chi_{AA}(t) = \chi_{AA}^{O}(t) + \chi_{AA}^{E}(t), \qquad (2.46)$$

where

$$\chi_{AA}^{E}(t) = \frac{\chi_{AA}(t) + \chi_{AA}(-t)}{2} \qquad (2.47)$$

and

$$\chi_{AA}^{O}(t) = \frac{\chi_{AA}(t) - \chi_{AA}(-t)}{2}. \qquad (2.48)$$

are the $\chi_{AA}(t)$ components even and odd upon time reversal, respectively. These functions are obviously defined for all times. We can now consider the Fourier transforms of $\chi_{AA}(t)$, customarily referred to as the generalized (ω-dependent) susceptibility $\hat{\chi}_{AA}(\omega)$, which reads as:

$$\hat{\chi}_{AA}(\omega) = \hat{\chi}_{AA}^{E}(\omega) + \hat{\chi}_{AA}^{0}(\omega) = \chi_{AA}'(\omega) + i\chi_{AA}''(\omega) \qquad (2.49)$$

Here we introduced the two spectral responses $\chi_{AA}'(\omega)$ and $\chi_{AA}''(\omega)$, respectively defined as the Fourier transform of $\chi_{AA}^{E}(t)$ and the imaginary part of the Fourier transform of $\chi_{AA}^{O}(t)$ changed of sign, respectively. These functions are both real and, respectively, even and odd in frequency. This stems from the circumstance that the Fourier transform of an odd real function is odd and imaginary while the one of an even real function of time is even and real.

At this stage, Eq. (2.42) can be rewritten in a way that makes it valid for all times. Namely:

$$2\chi_{AA}^{O}(t) = -\beta \frac{d}{dt} \langle \delta A(t)\delta A(0) \rangle.$$

The formula above reveals the odd character of this response function, at least in the classical case. In fact, from the stationarity property, the correlation function in the right hand side remains unchanged upon time reversal, i.e. upon $t \to -t$ change, while, obviously the operator d/dt changes sign. By Fourier transforming both members of the above equation and considering Eq. (2.49), one has:

$$\chi''_{AA}(\omega) = \frac{\beta\omega}{2} S_{\delta A}(\omega), \tag{2.50}$$

where it was used the relation $\chi''(\omega) = -i\hat{\chi}^O_{AA}(\omega)$. Furthermore, from the stationary character of equilibrium averages, it follows that:

$$S_{\delta A}(\omega) = \frac{1}{2\pi} \int_{-\infty}^{\infty} d\tau' \langle \delta A(t)\delta A(t+\tau') \rangle \exp(-i\omega\tau').$$

The equation above expresses in a more conventional form the main result of the fluctuation-dissipation theorem, by showing that $\hat{\chi}''_{AA}(\omega)$ is directly related to the Fourier transform of the time correlation of spontaneous fluctuations. This stresses the relevance of $\hat{\chi}''_{AA}(\omega)$ for *IXS* measurements, which, as demonstrated in the next sections, directly access the spectrum of the spontaneous fluctuations of microscopic density. Clearly, in the reciprocal space probed by *IXS*, the response function also depends on $\mathbf{Q}$.

Let us now consider a classical application for the memory function formalism, i.e. the one dealing with a damped harmonic oscillator of mass m subjected to a force $F(t)$. The time dependent displacement $x(t)$ of this system is described by the following differential equation:

$$\ddot{x}(t) + \gamma\dot{x}(t) + \Omega^2 x(t) = \frac{F(t)}{m}, \tag{2.51}$$

where γ is the damping coefficient, while $\Omega^2 = k/m$, with k being the force constant. The above equation can be represented in the frequency domain by Fourier transforming both sides, which yields:

$$\hat{x}(\omega) = \frac{\hat{F}(\omega)/m}{\Omega^2 - \omega^2 + i\gamma\omega}.$$

The search of an explicit expression for $x(t)$ can be framed in the context of the response function formalism, starting from a simple generalization of the cases previously encountered of stepwise or impulsive forces. For a force with a generic time dependence, one can write:

$$x(t) = \int_{-\infty}^{t} dt' \chi(t-t')F(t').$$

Since the displacement is here expressed as the convolution product between the force and the response function, its Fourier transform can be evaluated using the convolution theorem which yields:

$$\hat{x}(\omega) = \hat{\chi}(\omega)\hat{F}(\omega). \tag{2.52}$$

Fourier transforming both sides of the equation above, one obtains:

$$x(t) = \int_{-\infty}^{\infty} d\omega \hat{\chi}(\omega)\hat{F}(\omega)\exp(i\omega t) .$$

Let us now consider a force with the following time-dependence:

$$F(t) = F_0 \cos(\omega_0 t) = \frac{F_0}{2}[\exp(i\omega_0 t) + \exp(-i\omega_0 t)];$$

whose Fourier transform reads as:

$$\hat{F}(\omega) = \frac{F_0}{2}[\delta(\omega - \omega_0) + \delta(\omega + \omega_0)]$$

It can be now recognized that the power dissipated to keep the oscillator moving is $F(t)\dot{x}(t)$, where:

$$W(t) = F(t)\dot{x}(t) = F(t)\int_{-\infty}^{\infty} d\omega (i\omega)\hat{\chi}(\omega)\hat{F}(\omega)\exp(i\omega t)$$

$$= i\frac{F_0^2}{2}\omega_0[\hat{\chi}(\omega_0)\exp(-i\omega_0 t) + \hat{\chi}(-\omega_0)\exp(i\omega_0 t)]\cos(\omega_0 t).$$

Since such a power has an oscillating time dependence, it is more convenient to deal with the average power dissipated over a period of the force $T_0 = 2\pi/\omega_0$. Considering that $\hat{\chi}(\omega) = \chi'(\omega) + i\chi''(\omega)$, such an average dissipated power reads as:

$$\overline{W(t)} = \frac{F_0^2 \omega_0}{2}\chi''(\omega_0). \tag{2.53}$$

Which shows that the dissipated power uniquely depends on the imaginary part of the generalized susceptibility, which, as previously shown, is also directly connected to the power spectrum of fluctuations.

Thus far we did not consider the shape of the response function. The actual shape of the generalized susceptibility can be simply derived by comparing Eqs. 2.52 and 2.52, which yields:

$$\hat{\chi}(\omega) = \frac{1}{m}\left[\frac{1}{\omega^2 - \Omega^2 + i\gamma\omega}\right]$$

$$= \frac{1}{m}\left[\frac{\Omega^2 - \omega^2}{(\omega^2 - \Omega^2)^2 + \gamma^2\omega^2} + i\frac{\gamma\omega}{(\omega^2 - \Omega^2)^2 + \gamma^2\omega^2}\right]$$

The dissipated power can be easily obtained by inserting the imaginary part in Eq. 2.53, which yields:

$$\overline{W(t)} = \frac{F_0^2}{2m}\left[\frac{\gamma\omega_0^2}{(\omega_0^2 - \Omega^2)^2 + \gamma^2\omega_0^2}\right]. \tag{2.54}$$

Assuming, for instance, that the damped harmonic oscillator is underdamped, i.e. $\omega > \gamma/2$, the dissipated power exhibits clear maxima when the frequency of the forcing term ω_0 is equal to the natural frequency of the system Ω.

By using the result of the fluctuation dissipation theorem in Eq. (2.50) we can derive the spectrum associated to the variable $x(t)$, which in this merely classical treatment reads as as:

$$S_x(\omega) = \frac{k_B T}{m}\left[\frac{2\gamma}{(\omega^2 - \Omega^2)^2 + \gamma^2\omega^2}\right]. \tag{2.55}$$

Up to a normalization factor, a similar ω-profile will be encountered in the remainder of this book and it is customarily referred as the damped harmonic oscillator (DHO) model of the spectrum. In this context, one deals with the Q and ω-generalized susceptibility associated with the microscopic density fluctuation. The related spectral function is the dynamic structure factor, to be introduced in the last section of this chapter.

We finally mention that real and imaginary parts of the susceptibility are customarily referred to as the reactive and the dissipative parts of the susceptibility. In the specific contest of a spectral measurement these components are primarily connected to the shift and the damping of a collective excitation, respectively.

We end here our brief discussion of the formalism of response function. In the next section we start giving a closer look on the connections between the spectrum, or the spectral power of dynamic variables and the outcome of spectroscopy measurements.

2.6. The variable measured by a scattering measurement (Wang, 2012, Berne and Pecora, 1976)

Here we show how the outcome of a generic spectroscopic experiment relates to correlation functions of fluctuating variables. Let $A(t)$ be a sample stochastic property relevant to the scattering measurement; $A(t)$ can be expressed as a function of its Fourier components:

$$A_{T_0}(t) = \frac{1}{\sqrt{T_0}} \sum_n A_n \exp(i\omega_n t), \qquad (2.56)$$

where $\omega_n = 2\pi n/T_0$ and A_n are the frequency and amplitude of the nth Fourier component, respectively. The series development in the above equation can be interpreted assuming that the time dependence of $A(t)$ reproduces itself after the time of the measurement, that is, it is periodic with a period T_0. This implies that all the frequencies are integer multiples of the fundamental frequency $\omega_1 = 2\pi/T_0$.

As mentioned, the detector of a spectrometer measures a quantity proportional to the square value of $A_{T_0}(t)$ averaged over the time T_0:

$$\langle |A_{T_0}(t)|^2 \rangle = \frac{1}{T_0} \int_{-\frac{T_0}{2}}^{\frac{T_0}{2}} dt |A_{T_0}(t)|^2$$

$$= \frac{1}{T_0^2} \sum_{n,m} \int_{-\frac{T_0}{2}}^{\frac{T_0}{2}} dt A_n A_m^* \exp i(\omega_n - \omega_m)t \qquad (2.57)$$

We can now consider that the time lapse T_0 is much longer than the oscillation period of any of its Fourier components, i.e. $T_0 \gg 2/\omega_n$, for any n. In this long T_0 limit, one has:

$$\lim_{T_0 \to \infty} \frac{1}{T_0} \int_{-\frac{T_0}{2}}^{\frac{T_0}{2}} dt \exp i(\omega_n - \omega_m)t = \delta_{nm}. \qquad (2.58)$$

where δ_{nm} is the Kronecker delta, which is equal to 1 or 0, for $n = m$ and $n \neq m$, respectively.

The above expression of δ_{nm} can be used to simplify the integral on the right side of Eq. (2.57), thus eventually obtaining:

$$\langle |A_{T_0}|^2 \rangle = \frac{1}{T_0} \sum_n |A_n|^2. \qquad (2.59)$$

We now seek a more explicit expression for such an autocorrelation function at finite times. To do so, we can start from the defining the autocorrelation of $A(t)$:

$$\langle A^*(t)A(t+\tau')\rangle_{T_0} = \frac{1}{T_0} \int_{-\frac{T_0}{2}}^{\frac{T_0}{2}} dt A_{T_0}^*(t)A_{T_0}(t+\tau'), \qquad (2.60)$$

which can be further manipulated by replacing $A_{T_0}^*(t)$ and $A_{T_0}(t+\tau')$, with the corresponding Fourier expansions. This leads to:

$$\langle A^*(t)A(t+\tau')\rangle_{T_0} = \sum_{n,m} \frac{A_n A_m^*}{T_0} \int_{-\frac{T_0}{2}}^{\frac{T_0}{2}} dt \exp[i(\omega_n - \omega_m)t] \exp(i\omega_n\tau')$$

$$= \sum_n |A_n|^2 \exp(i\omega_n\tau'), \qquad (2.61)$$

By multiplying first and last member of this identity chain by $\exp(-i\omega_m\tau')$, performing their integral over the variable τ' between $-T_0/2$ and $T_0/2$, and using Eq. (2.58), one obtains:

$$|A_m|^2 = \int_{-\frac{T_0}{2}}^{\frac{T_0}{2}} d\tau' \langle A^*(t)A(t+\tau')\rangle_{T_0} \exp(-i\omega_m\tau')$$

$$= 2\pi S_A^{T_0}(\omega_m) \qquad (2.62)$$

The last equality defines the variable $S_A^{T_0}(\omega_m)$, as:

$$S_A^{T_0}(\omega_m) = \frac{1}{2\pi} \int_{-\frac{T_0}{2}}^{\frac{T_0}{2}} d\tau' \langle A^*(t)A(t+\tau')\rangle_{T_0} \exp(-i\omega_m\tau'), \qquad (2.63)$$

At this point upon inserting Eq. (2.61) into Eq. (2.58) one obtains:

$$\langle |A_{T_0}|^2\rangle = \frac{2\pi}{T_0} \sum_n S_A^{T_0}(\omega_n)$$

If we denote the constant difference between adjacent frequency values as $\Delta\omega_n = \omega_{n-1} - \omega_n = 2\pi/T_0$ and take the limit $T_0 \to \infty$, i.e. we assume that the frequency value defines a continuum, the above equation can be written as:

$$\lim_{T_0 \to \infty} \langle |A_{T_0}|^2 \rangle = \langle |A|^2 \rangle = \lim_{T_0 \to \infty} \sum_n \Delta\omega_n S_A^{T_0}(\omega_n)$$

$$= \int_{-\infty}^{\infty} S_A(\omega)d\omega, \qquad\qquad (2.64)$$

where:

$$S_A(\omega) = \frac{1}{2\pi} \int_{-\infty}^{\infty} d\tau' \langle A(t)A(t+\tau') \rangle \exp(-i\omega\tau') \qquad (2.65)$$

is customarily referred to as the spectral power — or, equivalently, the spectral density, or spectrum of the variable $A(t)$. As stated in Eq. (2.53), the quantity that the measurement ultimately detects is the first member of Eq. (2.60), which is thus equal to the integral of the spectral density.

Equation (2.65) is the main result of the so-called Wiener–Kintchine theorem (Khintchine, 1934, Wiener, 1930), which sets a link between the spectral power of a given variable $A(t)$ and the autocorrelation of such a variable. In particular, equations (2.60) and (2.61) set the link between the output of a measurement coupling with a given variable $A(t)$ and the spectral power of such a variable. Notice that, in the derivation above, we discarded the presence of frequency-filtering elements needed to implement the frequency analysis (scanning) of the output signal. In an inelastic spectrometer, two energy (or frequency) filtering elements, a monochromator and a frequency (energy) analyzer, are included downstream and upstream to the sample, respectively.

Let us assume that the downstream component (the monochromator) is used as a fixed filter of the frequency; the upstream one is used instead as a movable filter, i.e. as a filter whose passing frequency can be scanned within a given range, e.g. by rocking or heating a crystal. If the filtering elements are ideal, i.e. if they do not introduce any broadening of distortion of the input signal, their combined presence can be accounted for by introducing a $\delta(\omega)$-shaped "filter function" inside the integral in the last member of Eq. (2.60). It readily appears that the output signal detected while including these ideal filters is simply $S_A(\omega)$.

Let us now consider the case of a real energy analyzer. The distortion introduced by this element can be accounted for by introducing a "filtering function", usually referred to as the frequency (or energy) resolution function $R(\omega)$. This function can be defined as the spectral shape delivered by the analyzer when the input profile is a $\delta(\omega)$. As such, $R(\omega)$ provides a measure of the signal broadening and distortion induced by the analyzer.

If a monochromatic beam with a given frequency ω' impinges on the analyzer, the action of the latter can be accounted for by introducing a $R(\omega' - \omega)$ term, uniquely depending on the difference between the input and the (movable) analyzer's passing frequency ω. Expectedly $R(\omega' - \omega)$ tends to vanish upon an increase of $|\omega' - \omega|$, as frequencies too distant from the input one are filtered out. Upon including the factor $R(\omega' - \omega)$ into the integral of Eq. (2.60), the output of the measurement transforms into the convolution product between $R(\omega)$ and $S_A(\omega)$, as schematically represented in Figure 2-6.

In general, each optical element of the spectrometer contributes in a complicated fashion to the overall resolution of the instrument. As discussed in Chapter 4, the resolution profile is estimated by measuring the spectrum from an almost perfect elastic scatterer, i.e. using an input signal resembling a $\delta(\omega)$.

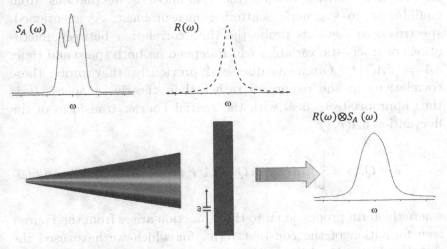

Figure 2-6: Schematic representation of the output of an analyzer filter (see text).

We now conclude this section with a few qualitative remarks on the link between autocorrelation function and spectrum.

As illustrated in the next paragraph — and consistently with what is discussed in the two previous ones — the variable of practical relevance for a spectroscopy measurement is the fluctuation of $A(t)$ from equilibrium, $\delta A(t) = A(t) - \langle A \rangle$.

One can easily recognize that, if $\delta A(t)$ varies rapidly with time, its spectrum is mainly dominated by high-frequency components; conversely, if it varies slowly, its spectral power prevalently spans the low-frequency window. Clearly, the adverbs "slowly" and "rapidly" only acquire a definite meaning in reference to the window of sensitivity of the considered technique and to the variable such a technique couples with; since high-resolution *IXS* probes the terahertz spectrum of density fluctuations, one might include sub-picosecond "in cage" atomic rattling and intramolecular vibrations among "rapid" dynamic processes and cooperative (structural) relaxations among "slow" ones.

2.7. Further remarks on the outcome of spectroscopic measurements

Let us now further discuss the relevance of fluctuations from equilibrium to a generic scattering measurement. As mentioned, spectroscopic methods probe the time correlation between fluctuations of stochastic variables which depend on both space and time: $\delta A = \delta A(r,t)$. Often, as discussed previously, they probe these correlations in the reciprocal rather than the direct space. It is thus appropriate to deal with the spatial Fourier transform of the fluctuation $\delta A(r,t)$:

$$\delta A(\boldsymbol{Q},t) = \left[\int_V d\boldsymbol{r} \exp(i\boldsymbol{Q} \cdot \boldsymbol{r}) A(\boldsymbol{r},t) \right] - \langle A \rangle \delta(\boldsymbol{Q}), \qquad (2.66)$$

where the term proportional to the δ-function arises from the Fourier transformation of the constant term, for which we have used the shorthand notation $\langle A \rangle = \langle A(\boldsymbol{r},t) \rangle$.

The expression embedded between square brackets in Eq. (2.64) is the Fourier transform of $A(r(t)$, which thus reads as:

$$A(\boldsymbol{Q},t) = \delta A(\boldsymbol{Q},t) + \langle A \rangle \delta(\boldsymbol{Q})$$

$$= \int_V d\boldsymbol{r} \exp(i\boldsymbol{Q} \cdot \boldsymbol{r}) A(\boldsymbol{r},t). \tag{2.67}$$

Considering the equation above, the spectral power of this variable is given by:

$$S_A(\boldsymbol{Q},\omega) = \frac{1}{2\pi} \int_{-\infty}^{\infty} dt \langle \delta A(\boldsymbol{Q},t) \delta A^*(\boldsymbol{Q},0) \rangle \exp(-i\omega t)$$

$$+ \langle A \rangle^2 \delta(\omega) \delta(\boldsymbol{Q}), \tag{2.68}$$

which can be written as:

$$S_A(\boldsymbol{Q},\omega) = S_{\delta A}(\boldsymbol{Q},\omega) + \langle A \rangle^2 \delta(\omega) \delta(\boldsymbol{Q}). \tag{2.69}$$

The term $\propto \delta(\omega)\delta(\boldsymbol{Q})$ is non-vanishing only at $\omega = 0$ and $\boldsymbol{Q} = 0$. Considering the link between conjugate Fourier variables, we conclude that this contribution results from an average of the system response for infinitely long times and distances. Here the sample appears as a continuous and homogeneous system at equilibrium.

Notice that this term vanishes unless the probe beam has no exchanges of energy ($\hbar\omega = 0$) or momentum ($\hbar\boldsymbol{Q} = 0$) with the sample.

This contribution is not only of limited interest, it is also practically inaccessible by a scattering measurement. In fact, the $\delta(\boldsymbol{Q})$ function is non-vanishing for $\boldsymbol{Q} = 0$ only i.e. in the forward direction, where the dominating intensity comes from the transmitted beam, and typically is high enough to saturate or burn scattering detectors.

Therefore, it can be assumed that this identification holds validity:

$$S_A(\boldsymbol{Q},\omega) \equiv S_{\delta A}(\boldsymbol{Q},\omega). \tag{2.70}$$

It follows that, at finite $\boldsymbol{Q}$'s, the spectral density measured by a probe coupling with some variable A, essentially coincides with the spectral density of its fluctuations from equilibrium. This corroborates our previous statement (see, e.g., Chapter 1) that the scattering mainly arises from fluctuations in the scattering medium. These, of course, may be time-dependent and reflect the internal dynamics of the system, or space-dependent and relate, instead, to static, or structural inhomogeneities of the sample.

A reasonable assumption made throughout this book is that for isotropic systems as liquids and gases, the relevant variable is not the vector $\boldsymbol{Q}$, but rather its modulus $Q = |\boldsymbol{Q}|$, which authorizes to drop the vector (bold) notation when expressing the wavevector dependence of various spectral functions.

2.8. A simple method to derive some prototypical spectral shapes

From its very definition, the spectrum of the variable $A(t)$ is the Fourier transform of the autocorrelation function $C_{AA}(t)$; thus, it reads as:

$$S_A(\omega) = \frac{1}{2\pi} \int_{-\infty}^{\infty} dt \exp(-i\omega t) C_{AA}(t). \qquad (2.71)$$

Using the time evenness of the autocorrelation function in a classical system, the above integral can be split as:

$$S_A(\omega) = \frac{1}{2\pi} \int_{-\infty}^{0} dt \exp(-i\omega t) C_{AA}(t) + \frac{1}{2\pi} \int_{0}^{\infty} dt \exp(-i\omega t) C_{AA}(t).$$

At this point, it is useful to replace, in the first integral, the integration variable t with $-t$, thus obtaining:

$$S_A(\omega) = \frac{1}{2\pi} \int_{0}^{\infty} dt \exp(i\omega t) C_{AA}(-t) + \frac{1}{2\pi} \int_{0}^{\infty} dt \exp(-i\omega t) C_{AA}(t)$$

$$= \frac{1}{2\pi} \int_{0}^{\infty} dt \exp(i\omega t) C_{AA}^*(t) + \frac{1}{2\pi} \int_{0}^{\infty} dt \exp(-i\omega t) C_{AA}(t),$$

where we have used the property of the autocorrelation function $C_{AA}(-t) = C_{AA}^*(t)$.[2] Furthermore, from the formula

---

[2]See footnote in section 2.2.

$\exp(i\omega t)C_{AA}^*(t) + \exp(-i\omega t)C_{AA}(t) = 2Re[\exp(-i\omega t)C_{AA}(t)]$, it follows that:

$$S_A(\omega) = \frac{1}{\pi} Re \left[\int_0^\infty dt \exp(-i\omega t)C_{AA}(t) \right].$$

Notice that the expression between square brackets is the Laplace transform of $C_{AA}(t)$, $\widetilde{C}_{AA}(z)$ (where $z = i\omega$). Hence we will often use the following notation for various spectral function:

$$S_A(\omega) = \frac{1}{\pi} Re[\widetilde{C}_{AA}(z = i\omega)]. \tag{2.72}$$

For instance, one can use this formula to compute the spectral density from a collective excitation having frequency ω_0 and damping Γ:

$$A_w(t) = A_w \cos(\omega_0 t) \exp(-\Gamma t),$$

where the subscript "w" hints at the wave character of the excitation. Using Eq. (2.72) one can readily derive the following spectral shape:

$$\begin{aligned}
S_A(\omega) &= \frac{A_w^2}{\pi} Re \left[\int_0^\infty dt \cos[\omega_0 t] \exp(-\Gamma t) \exp(i\omega t) \right] \\
&= \frac{A_w^2}{\pi} Re \left\{ \frac{1}{[i(\omega - \omega_0) - \Gamma]} + \frac{1}{[i(\omega + \omega_0) - \Gamma]} \right\} \\
&= \frac{A_w^2}{\pi} \left\{ \frac{\Gamma}{[(\omega - \omega_0)^2 + \Gamma^2]} + \frac{\Gamma}{[(\omega - \omega_0)^2 + \Gamma^2]} \right\},
\end{aligned}$$

which consists of two Lorentzian profiles centred at $\pm\omega_0$ and having half-width at half maximum Γ. It appears that a mode having a finite frequency (ω_0) has a spectral density featuring two inelastic peaks symmetrically shifted from the elastic position, $\omega = 0$.

Let us now consider a collective mode with no oscillatory behaviour, for instance, connected to thermal diffusion or to a relaxation process. Its time dependence can be written in the form:

$$A(t) = A_D \exp(-\Delta t).$$

It is immediately verified that the spectrum, in this case, reduces to a Lorentzian centred at the origin $\omega = 0$ and having half-width at half maximum equal to the coefficient Δ, which coincides with the inverse of the time-decay of the excitation.

If a given mode has a non-vanishing dominant frequency, i.e. if it corresponds to a spectral feature shifted from the elastic position, such a mode will be hereafter referred to as inelastic. Conversely, if its spectral contribution gathers at the $\omega = 0$ (elastic position) to within a relatively narrow frequency distribution, this mode will be often referred to as quasi-elastic.

2.9.   Some variables, correlation and spectral functions of interest (Balucani and Zoppi, 1994, Hansen and McDonald, 2006)

As mentioned before and thoroughly discussed in the next chapter, the *IXS* technique measures the spectrum of the microscopic density fluctuation. Let us briefly discuss how this key variable can be expressed in analytical form.

A suitable expression must mirror the highly discontinuous mass distribution at microscopic scales; in particular, it is reasonable to assume that the microscopic density associated to the single atom vanishes everywhere except at the positions, $\boldsymbol{R}_i(t)$, of the ith nucleus.

For a system of N atoms, the microscopic density can thus be appropriately represented as the sum of N Dirac's δ-functions:

$$n(\boldsymbol{r}, t) = \sum_{j=1}^{N} \delta[\boldsymbol{r} - \boldsymbol{R}_j(t)]. \qquad (2.73)$$

The jth term of the sum is the "self", or single particle, density:

$$n_s(\boldsymbol{r}, t) = \delta[\boldsymbol{r} - \boldsymbol{R}_j(t)]. \qquad (2.74)$$

The actual identification of $n(\boldsymbol{r}, t)$ with a microscopic number density becomes more evident after recognising that its integral over a finite volume V, $\int_V d\boldsymbol{r} n(\boldsymbol{r}, t)$, is equal to the number of atoms contained in such a volume. In the continuous, or macroscopic limit, corresponding to $Q \to 0$, the number density becomes a globally averaged property of the entire system, thus no longer depending on its local (space and time) coordinates, i.e., in this regime, the identity $n(\boldsymbol{r}, t) \equiv n$ holds validity. Indeed, here $n \int_V d\boldsymbol{r} = N$ or equivalently $n = N/V$, which is consistent with the definition of the macroscopic number density $n = \langle n(\mathbf{r}, t) \rangle$.

Considering that a spectroscopic experiment probes the reciprocal space, it is convenient to introduce the spatial Fourier transform of the microscopic density:

$$n(\mathbf{Q}, t) = \sum_{j=1}^{N} \int_{V} d\mathbf{r} \exp(i\mathbf{Q} \cdot \mathbf{r}) \delta[\mathbf{r} - \mathbf{R}_j(t)] = \sum_{j=1}^{N} \exp[i\mathbf{Q} \cdot \mathbf{R}_j(t)].$$

(2.75)

It is instructive to have a closer look at the time derivative of this variable:

$$\dot{n}(\mathbf{Q}, t) = \frac{dn(\mathbf{Q}, t)}{dt} = \sum_{j=1}^{N} i\mathbf{Q} \cdot \mathbf{v}_j(t) \exp[i\mathbf{Q} \cdot \mathbf{R}_j(t)]. \qquad (2.76)$$

Here $\mathbf{v}_j(t) = d\mathbf{R}_j(t)/dt$ is the velocity of the jth atom. From the above equation, it follows that $\lim_{\mathbf{Q} \to 0} \dot{n}(\mathbf{Q}, t) = 0$, which implies that the number density in the macroscopic limit does not change with time, which mirrors the conservation of the total number, N, of atoms in the system.

As perhaps clear from the previous paragraphs, the crucial dynamic variable for spectroscopic experiments is:

$$\delta n(\mathbf{r}, t) = \sum_{j=1}^{N} \delta[\mathbf{r} - \mathbf{R}_j(t)] - n, \qquad (2.77)$$

which represents the fluctuation of the microscopic density from its average equilibrium value, n.

The Fourier transform of the above variable is:

$$\delta n(\mathbf{Q}, t) = \sum_{j=1}^{N} \exp[i\mathbf{Q} \cdot \mathbf{R}_j(t)] - n\delta(\mathbf{Q}). \qquad (2.78)$$

We can now introduce a few relevant correlation functions involving dynamic variables of the system. The first is a time-dependent correlation which, rather than a single atom, involves a pair of atoms labelled by the indices k and j and it is called the van Hove function.

To introduce this correlation function within a merely classical treatment, one can start from fixing the origin at the position of the tagged kth atom, whose *self* microscopic density can be written as $\delta[r' - R_k(0)]$. Regardless of the location of the kth atom for $t = 0$, we assume that, at the time t, the jth atom locates at a distance $R_j(t) = r + r'$ from the origin and, its *self* microscopic density reads as $\delta[r + r' - R_j(t)]$.

We can now define a microscopic density associated with the pair as the product:

$$\delta[r + r' - R_j(t)]\delta[r' - R_k(0)]. \qquad (2.79)$$

Since we are considering homogeneous many-body systems, the property we are attempting to build in must not depend on the position of the origin; therefore, the product above should be integrated over all possible values of r'. Therefore, we should rather deal with

$$\int_V dr' \delta[r + r' - R_j(t)]\delta[r' - R_k(0)]. \qquad (2.80)$$

For each tagged kth atom, one must execute a sum over all jth atoms; moreover, the results must be averaged over all possible N atoms of the system, so that eventually the average assumes the form $1/N \sum_{j=1}^{N} \sum_{k=1}^{N} \langle \cdots \rangle$. This yields the so-called van Hove correlation function:

$$G(r, t) = \frac{1}{N} \sum_{jk=1}^{N} \int_V dr' \langle \delta[r + r' - R_j(t)]\delta[r' - R_k(0)]\rangle, \qquad (2.81)$$

with the shorthand notation $\sum_{jk=1}^{N} \cdots = \sum_{j=1}^{N} \sum_{k=1}^{N} \cdots$; as usual, the average is performed as a thermal average on the sample at equilibrium. The inclusion of the normalization factor $1/N$ ensures that the van Hove function is equal to 1 for $t = 0$, as quickly verified by considering the integral over the volume of the whole sample.

The formula above can be rearranged while considering the Fourier components of the microscopic density, thus eventually obtaining:

$$G(r, t) = \frac{1}{N} \sum_{j=1}^{N} \sum_{k=1}^{N} \langle \delta[r + R_k(0) - R_j(t)]\rangle. \qquad (2.82)$$

Notice that for a system as a simple liquid, which is invariant not only for translations (homogeneous) but also for rotations (isotropic), the van Hove function does not depend on the direction of r but only on its modulus $|r|$.

At this stage, one can rewrite the above equation as the sum of two terms involving different correlations:

(1) The terms with $j = k$ are "self" autocorrelations which involve the same atom at different times and
(2) The terms with $j \neq k$ are "distinct" autocorrelations which, instead, involve different atoms at different times.

It is customary to write correlation functions and related spectral densities as the sum of these *self* and distinct terms. Hence, for instance, $G(r,t) = G_s(r,t) + G_d(r,t)$, where:

$$G_s(r,t) = \frac{1}{N} \sum_{j=1}^{N} \langle \delta[r + R_j(0) - R_j(t)] \rangle \qquad (2.83)$$

and

$$G_d(r,t) = \frac{1}{N} \sum_{j=1}^{N} \sum_{k \neq j=1}^{N} \langle \delta[r + R_j(0) - R_k(t)] \rangle \qquad (2.84)$$

are the self and the distinct parts of the van Hove function, respectively. It is useful to consider the $t = 0$ value of the Van Hove function in a homogeneous and isotropic system, i.e., as mentioned, a system which is invariant for translations and rotation respectively. One has:

$$G(r,0) = \left\langle \frac{1}{N} \sum_{j=1}^{N} \sum_{k=1}^{N} \delta[r + R_k(0) - R_j(0)] \right\rangle = \delta(r) + ng(r).$$
$$(2.85)$$

In the rightmost side of this identity chain, the δ-function arises from the self ($j = k$) terms of the sum, which involves the position of a single atom; the part proportional to $g(r)$ originates instead from pairs of distinct atoms, for which $j \neq k$. The variable $g(r)$ is called the pair distribution function, and is accessible by both numerical computations and diffraction measurements. From the physical point

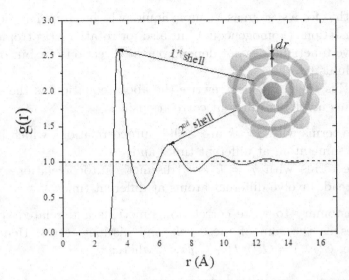

Figure 2-7:   The pair distribution function of a liquid. The included sketch portrays the contribution of successive atomic shells.

of view, $g(r)$ represents the probability distribution that an atom of the fluid is found at a distance r from the (different) tagged atom sitting in the origin (see Figure 2-7). Notice that at long distances, $G(\boldsymbol{r}, 0)$ tends to the (macroscopic) number density n, which, from the equation above, implies that $g(r)$ tends to 1 in the same limit.

The integral of $ng(r)$ over a volume V is equal to the number of atoms inside such a volume. As evident from Figure 2-8, the number of atoms at a distance r from a tagged one systematically increases with increasingly damped oscillations, eventually joining the expected r^3 macroscopic trend until the value $n(r) = N - 1$ is reached.

The Fourier transform of the van Hove function:

$$F(Q,t) = \frac{1}{N} \int_V \langle \delta n(\boldsymbol{r},t) \delta n^*(\boldsymbol{r},0) \rangle \exp(i\boldsymbol{Q} \cdot \boldsymbol{r}) d\boldsymbol{r}, \qquad (2.86)$$

is usually referred to as the intermediate scattering function. It can be easily verified that the above function can be also expressed as $F(Q,t) = \frac{1}{N} \langle \delta n(\boldsymbol{Q},t) \delta n^*(\boldsymbol{Q},0) \rangle$ or $F(Q,t) = \frac{1}{N} \langle \delta n(\boldsymbol{Q},t) \delta n(-\boldsymbol{Q},0) \rangle$

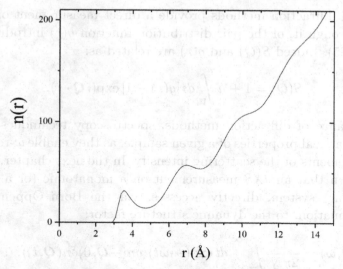

Figure 2-8: The typical profile, in a liquid system, of the number of atoms $n(r) = n \int_0^r 4\pi r^2 g(r)dr$ within a variable distance r from a tagged particle.

since:

$$\delta n^*(\boldsymbol{Q}, t) = \left\{ \sum_{j=1}^{N} \exp[i\boldsymbol{Q} \cdot \boldsymbol{R}_j(t)] \right\}^*$$

$$= \sum_{j=1}^{N} \exp[-i\boldsymbol{Q} \cdot \boldsymbol{R}_j(t)] = \delta n(-\boldsymbol{Q}, t).$$

Notice that, due to the isotropy of the fluid, $F(Q, t)$, as well other microscopic properties or response functions, depends on the amplitude of $\boldsymbol{Q}$, but not on its direction. At this stage, it is useful to introduce the static structure factor:

$$S(Q) = F(Q, t = 0) = \frac{1}{N} \int_V \langle \delta n(\boldsymbol{r}, 0) \delta n^*(\boldsymbol{r}, 0) \rangle \exp(i\boldsymbol{Q} \cdot \boldsymbol{r})d\boldsymbol{r},$$

$$(2.87)$$

which is the spatial Fourier transform of the "static" $(t = 0)$ density correlation function, thus conveying direct insight into the structural properties of the fluid. Most importantly, both X-Ray or neutron diffraction methods provide a direct measurement of $S(Q)$ and, through it, of the pair distribution function $g(r)$ introduced in Eq. (2.85). Indeed $S(Q)$ and $g(r)$ are related as:

$$S(Q) = 1 + n \int_V d\boldsymbol{r}[g(r) - 1] \exp(i\boldsymbol{Q} \cdot \boldsymbol{r}). \qquad (2.88)$$

At variance of diffraction methods, spectroscopy techniques probe the dynamical properties of a given sample, as they enable ω-resolved measurements of the scattering intensity. In the next chapter, it will be shown that an *IXS* measurement on a monatomic (or a simple molecular) system, directly accesses, via the Born–Oppenheimer approximation, to the dynamic structure factor:

$$S(Q,\omega) = \frac{1}{2\pi N} \int_{-\infty}^{+\infty} dt \exp(-i\omega t) \langle \delta n(-\boldsymbol{Q}, 0)\delta n(\boldsymbol{Q}, t) \rangle. \qquad (2.89)$$

From this definition, it appears that the dynamic structure factor is the Fourier transform of the intermediate scattering function introduced in Eq. (2.86).

Some correlation functions cannot be measured directly, but they can be computed via Molecular Dynamics (MD) computer simulations; these are, for instance, those involving the variable microscopic (density of) velocity:

$$v(\boldsymbol{r}, t) = \sum_{j=1}^{N} \boldsymbol{v}_j(t)\delta[\boldsymbol{r} - \boldsymbol{R}_j(t)],$$

or its Fourier transform, which is often referred to as the microscopic flow, or the current:

$$\boldsymbol{J}(\boldsymbol{Q}, t) = \sum_{j=1}^{N} \boldsymbol{v}_j(t) \exp[i\boldsymbol{Q} \cdot \boldsymbol{R}_j(t)]. \qquad (2.90)$$

This vector variable can be represented as the sum of its projections either along $\hat{\boldsymbol{Q}}$, or in the plane orthogonal to it:

$$\boldsymbol{J}(\boldsymbol{Q}, t) = \hat{\boldsymbol{Q}}\hat{\boldsymbol{Q}} \cdot \boldsymbol{J}(\boldsymbol{Q}, t) + (I - \hat{\boldsymbol{Q}}\hat{\boldsymbol{Q}}) \cdot \boldsymbol{J}(\boldsymbol{Q}, t) - \boldsymbol{J}_L(\boldsymbol{Q}, t) + \boldsymbol{J}_T(\boldsymbol{Q}, t),$$

where the suffixes "L" and "T" here label longitudinal and transverse projections of the current $\boldsymbol{J}(\boldsymbol{Q},t)$, respectively.

In the macroscopic limit and for a system not subjected to external forces, the variable current is conserved, i.e. it is constant in time, as immediately verified by computing its time derivative and considering Newton's law of momentum conservation.

If one, for instance, sets the vector $\boldsymbol{Q}$ parallel to the z-axis, the Fourier transforms of the autocorrelation functions of transverse and longitudinal currents read as:

$$C_L(Q,\omega) = \frac{1}{2\pi N} \int_{-\infty}^{\infty} dt \langle J_z(Q,t) J_z^*(Q,0) \rangle \exp(-i\omega t) \qquad (2.91)$$

and

$$C_T(Q,\omega) = \frac{1}{2\pi N} \int_{-\infty}^{\infty} dt \langle J_x(Q,t) J_x^*(Q,0) \rangle \exp(-i\omega t), \qquad (2.92)$$

respectively, with:

$$J_z(Q,t) = \frac{1}{2\pi} \sum_{j=1}^{N} v_j^z(t) \exp[iQz_j(t)]$$

and

$$J_x(Q,t) = \frac{1}{2\pi} \sum_{j=1}^{N} v_j^x(t) \exp[iQx_j(t)].$$

Here the suffix labels the projection of the vector on the corresponding axis. Notice that, for the isotropy of the system, the tranverse current in Eq. (2.90), is unaltered if the component along x are replaced by the ones along the other transverse coordinate y.

It can be shown that the longitudinal current spectrum is directly related to the dynamic structure factor through:

$$C_L(Q,\omega) = \frac{\omega^2}{Q^2} S(Q,\omega). \qquad (2.93)$$

While searching for collective modes dominating the dynamics of a system, it is often convenient to deal with $C_L(Q,\omega)$ rather than $S(Q,\omega)$, as the former spectral function exhibits at least a peak at a finite frequency. In fact, due to the presence of the ω^2/Q^2 pre-factor $C_L(Q,0) = 0$; furthermore, upon increasing ω, $S(Q,\omega)$ vanishes more rapidly than ω^2, which implies that also $C_L(Q, \omega \to \infty) = 0$. Since $C_L(Q,\omega)$ is a positive definite function — unless identically equal to 0 — it must have at least one maximum at intermediate frequencies.

The frequency, Ω, of such a peak is often identified as the dominant acoustic frequency. This interpretation becomes clearly elusive when the acoustic excitation is overdamped, i.e. when the damping of the sound mode exceeds the acoustic frequency. In fact, under these conditions, no peak exists in $S(Q,\omega)$, while a $C_L(Q,\omega)$ maximum still persists.

For instance, we will discuss explicitly some cases in which collective excitations become overdamped around the Q-position maximum of the first sharp maximum of $S(Q)$, and, correspondingly, a so-called propagation gap emerges in the dispersion curve. Clearly, this scenario makes questionable the identification between the dominant collective modes of either $S(Q,\omega)$ or $C_L(Q,\omega)$. Conversely, such identification becomes rigorous in the $Q \to 0$ limit, where $\Omega = \omega_s = c_s Q$ and the damping of the acoustic wave is much smaller than its frequency.

Appendix 2A: The Kramers-Kroenig relations

Let us look at the consequence of the causality principle for the response function. We will consider here the case in which $A = B$, so we can omit the subscript in the response function, by using the notation $\chi(t) = \chi_{AA}(t)$. In particular, here we deal with the generalized susceptibility $\hat{\chi}(\omega)$. Since, for the causality principle, $\chi(t) = 0, \forall t < 0$, the generalized susceptibility can also be written as:

$$\hat{\chi}(\omega) = \frac{1}{2\pi} \int_{-\infty}^{\infty} dt \exp(-i\omega t)\theta(t)\chi(t)$$

Where $\theta(t)$ is the Heavyside step function.

It can be recognized that the expression in the righthand side is the Fourier transform of the product $\chi(t)\theta(t)$. We can thus apply the convolution theorem according to which:

$$FT[\chi(t)\theta(t)] = \hat{\theta}(\omega) \otimes \hat{\chi}(\omega),$$

where in the lefthand side of the equation, $FT[\chi(t)\theta(t)]$ indicates the Fourier transform of the product $\chi(t)\theta(t)$. In particular, the Fourier transform of the Heaviside function, $\hat{\theta}(\omega)$, can be computed by considering that $\theta(t) = \lim_{\varepsilon \to 0} \exp(-\varepsilon t)$ for $t > 0$ and equal to 0 for $t < 0$. Therefore, its Fourier transform reads as:

$$\hat{\theta}(\omega) = \frac{1}{2\pi} \lim_{\varepsilon \to 0} \int_0^\infty dt \exp(-\varepsilon - i\omega)t = \frac{1}{2\pi} \lim_{\varepsilon \to 0} \int_0^\infty \exp[-i(\omega - i\varepsilon)t]$$

$$= \frac{1}{2\pi i} \lim_{\varepsilon \to 0} \frac{1}{\omega - i\varepsilon} = \frac{1}{2\pi i} [\mathcal{P}\frac{1}{\omega} + \pi i \delta(\omega)].$$

Hence:

$$\hat{\chi}(\omega) = \frac{1}{2\pi i} \hat{\chi}(\omega) \otimes \left(\mathcal{P}\frac{1}{\omega} + \pi i \delta(\omega) \right), \qquad (2.94)$$

where $\mathcal{P}$ indicates the Cauchy principal value. At this stage, one can write explicitly the convolution product:

$$\hat{\chi}(\omega) = \frac{1}{2\pi} \left[\mathcal{P} \int_{-\infty}^\infty d\omega' \frac{\hat{\chi}(\omega')}{i(\omega' - \omega)} + \pi \hat{\chi}(\omega) \right] \implies$$

$$\hat{\chi}(\omega) = \frac{1}{\pi i} \mathcal{P} \int_{-\infty}^\infty d\omega' \frac{\hat{\chi}(\omega')}{(\omega' - \omega)} . \qquad (2.95)$$

As a further development, we can express the generalized susceptibility in terms of its real and imaginary parts: $\chi(\omega) = \chi'(\omega) + i\chi''(\omega)$. It appears that the introduction of Heaviside function, i.e. the superimposition of the causality principle, ultimately causes the imaginary and real part to be entangled to each other, as readily follows from the equation below:

$$\chi'(\omega) + i\chi''(\omega) = \frac{1}{\pi} \left[\mathcal{P} \int_{-\infty}^\infty \frac{\chi''(\omega')}{\omega' - \omega} d\omega' - i\mathcal{P} \int_{-\infty}^\infty \frac{\chi'(\omega')}{\omega' - \omega} d\omega' \right].$$

$$(2.96)$$

Indeed, by comparing real and imaginary parts in the above equation, one obtains:

$$\chi'(\omega) = \frac{1}{\pi}\mathcal{P}\int_{-\infty}^{\infty} d\omega' \frac{\chi''(\omega')}{\omega' - \omega} \tag{2.97}$$

$$\chi''(\omega) = -\frac{1}{\pi}\mathcal{P}\int_{-\infty}^{\infty} d\omega' \frac{\chi'(\omega')}{\omega' - \omega}, \tag{2.98}$$

The formulas above are usually referred to as the Kramers-Kronig relations. They show that, not only real and imaginary parts of the susceptivity are related to each other, but once one of them is known, the other can be generated from it.

References

Balucani, U. & Zoppi, M. 1994. *Dynamics of the Liquid State*, Oxford, Clarendon Press.

Batista, V. Introduction to Statistical Mechanics, Lecture Notes CHEM 530b, see website: http://ursula.chem.yale.edu/~batista/classes/vaa/vaa.pdf

Berne, B. J. & Pecora, R. 1976. *Dynamic Light Scattering*, New York, Wiley.

Boon, J. P. & Yip, S. 1980. Molecular Hydrodynamics, Dover, New York.

Callen, H. B. & Welton, T. A. 1951. *Phys. Rev.*, **83**, 34.

Hansen, J.-P. & McDonald, I. R. 2006. Chapter 7 — Time-Dependent Correlation and Response Functions. *In:* Hansen, J.-P. & McDonald, I. R. (eds.) *Theory of Simple Liquids (Third Edition)*. Burlington: Academic Press.

Khintchine, A. 1934. *Math. Ann.*, **109**, 604.

Kubo, R. 1966. *Rep. Prog. Phys.*, **29**, 255.

MacKintosh, F. *Lecture Notes on Linear Response Theory* [Online]. Vrije Universiteit. Available: https://web.science.uu.nl/DRSTP/Postgr.courses/SPTCM/2011/MacKintosh.pdf [Accessed].

Onsager, L. 1931a. *Phys. Rev.*, **37**, 405.

Onsager, L. 1931b. *Phys. Rev.*, **38**, 2265.

Wang, C. H. 2012. *Spectroscopy of Condensed Media: Dynamics of Molecular Interactions*, Elsevier.

Wiener, N. 1930. *Acta Mat.*, **55**, 117.

Yip, S. Microscopic Theory of Transport, MIT Lecture Notes, Lecture 3 (2003), see website: https://ocw.mit.edu/courses/nuclear-engineering/22-103-microscopic-theory-of-transport-fall-2003/lecture-notes/

Chapter 3

The *IXS* technique

After discussing in the past chapters how the spectral response relates to correlation functions of fluctuating variables of a fluid at equilibrium, the general purpose of this chapter is helping the reader to become familiar with relevant theoretical aspects of Inelastic X-Ray Scattering. Specifically, the main aim of the following pages is to illustrate the analytical derivation of the scattering cross-section and to demonstrate its direct link to the spectrum of density fluctuation of the sample.

3.1. Generalities on an *IXS* experiment

All existing *IXS* spectrometers use an undulator in the accumulation ring of a synchrotron as a source for the X-Ray beam. A detailed description of synchrotron research facilities goes beyond the scope of this book, and the interested reader is referred to Ref. (Mobilio *et al.*, 2016).

An undulator generates an X-Ray photon beam whose propagating direction is defined within a small angular divergence by a combination of optical elements. This collimation is essential to determine precisely the direction of the incident phonon wavevector k_i, whose amplitude $|k_i| = (2\pi/\hbar c)E_i$ is instead fixed by the monochromator, through the selection of the incident beam energy $E_i = \hbar\omega_i$. Here c and $\hbar = h/2\pi$ are the speed of light in vacuum and the reduced Plank constant, respectively. Also, incident photons have a well-defined polarization identified by the unit vector $\hat{\varepsilon}_i$ which, for undulator-emitted beams, lies in the plane of the synchrotron orbit.

The photon beam so shaped is then focused on the sample, typically by a single or a combination of mirrors and, sometimes, a compound of X-Ray refractive lenses. Although the focusing makes the photon trajectories convergent, thereby increasing the spread in the k_i direction, it is often required to reduce the focal spot to match the usually small area accepted by the downstream optics, or when measuring the scattering from small samples. Eventually, $\dot{N}_i$ photons per second impinge on the sample.

Due to the interaction between the incident photon beam and the target sample, some photons are scattered in all directions and only a small photon rate, $\dot{N}_i$, is deviated by a scattering angle 2θ to within the solid angle $\Delta\Omega = (A/r^2)$ covered by the detector, which has sensitive area A and distance r from the centre of the sample. Notice that the orientation of isotropic, non-crystalline samples, like the ones of interest for this book, is a parameter irrelevant to the scattering geometry, which is entirely defined by the scattering angle 2θ.[1]

The photons deflected by an angle 2θ pass through the analyser and the collimating units, whose combined effect is to select their energy $E_f = \hbar\omega_f$, the direction of the final wavevector k_f and its amplitude $|k_f| = (2\pi/\hbar c)E_f$, as well as their polarization, identified by the versor $\hat{\varepsilon}_f$.

The intensity of the scattered beam so shaped is ultimately measured by the detector.

During the scattering event, the impinging photon beam and the target sample exchange energy $\hbar\omega$ and momentum $\hbar Q$, typically within a time-lapse much shorter than any timescale relevant to the experiment. Indeed, the scattering process resembles a collisional event in which the total energy and momentum are conserved, and

---

[1]In general a direction in the three dimensional space is defined by two angles, but we only need one of them, 2θ, to describe the scattering geometry. The reason is that the scattering process is essentially probed in two dimensions i.e. in the plane defined by incident and scattered beams directions, to within the small solid angle intercepted by the detector.

their conservation determines the energy and momentum exchanged between sample and probe, which are respectively equal to $\hbar\omega = \hbar(\omega_i - \omega_f)$ and $\hbar(\mathbf{k}_i - \mathbf{k}_f)$. Notice that a positive $\hbar\omega$ value corresponds to an energy loss suffered from the photons after the scattering process.

Hereafter, we will refer to the whole instrument enabling the measurement of the *IXS* intensity as the *IXS* spectrometer. The actual complexity of this instrument may vary as required by the physical problem under investigation and the corresponding energy and momentum resolution needs. An example of a real *IXS* spectrometer is discussed at the end of Chapter 4.

Regardless of the specific layout, the general aim of an *IXS* spectrometer is to investigate the properties of the target sample through the detection and energy analysis of X-Ray photons which have interacted with it in the scattering event. The purpose of the next few sections is to derive a formal expression for the *IXS* scattering cross-section and ultimately show how this quantity is linked to the spectrum of density fluctuations, or dynamic structure factor, of the sample.

3.2. Introducing the differential cross-section (Sakurai and Commins, 1995, Fowler, 2003)

A simple formalization of a scattering problem is based upon the time-independent Schrödinger's equation for a quantum particle (schematized as a plane wave) impinging on a region of space where a localized potential $V(\mathbf{r})$ exists. Analytical details of this treatment are illustrated in Appendix 3A. As a result, this time-independent treatment shows that, given an impinging plane wave $\exp(i\mathbf{k}_i \cdot \mathbf{r})$, in first Born approximation, the wave function describing the scattered radiation at long distances from the target sample assumes the form:

$$u(\mathbf{r}) = \exp(i\mathbf{k}_i \cdot \mathbf{r}) + f_B(\mathbf{k}_i, \mathbf{k}_f)\frac{\exp(ikr)}{r}. \qquad (3.1)$$

Here the subscript "B" alludes to the Born approximation value of the parameter $f_B(k_i, k_f)$.[2] The latter has the dimension of a length and is usually referred to as the scattering length.

In Appendix 3A the Eq. (3.1) is derived for a massive quantum particle, to be identified, e.g. with a neutron, or an electron; however, a formally similar result can be obtained by using the Maxwell equations to achieve a non-relativistic and time-independent derivation of the scattering of electromagnetic radiation from an inhomogeneous medium (Born and Wolf, 2013). In this description, the interaction between the incident wave and the medium is linear, and the properties of the medium are described by macroscopic coefficients which do not change with time.

In summary, from Eq. (3.1) it can be concluded that, by virtue of the scattering event, the plane wave impinging on the sample is partly re-radiated into a spherical wave, which is one of the possible manifestations of the Huygens-Fresnel principle.

Although somehow simplified, the time-independent description discussed in Appendix 3A provides a solid ground to introduce some physical variables of interest in the successive, time-dependent, perturbative treatment of the inelastic X-Ray scattering.

The differential cross-section.

Following the same reasoning pattern illustrated in the textbook of Sakurai (Sakurai 1995), we can try here to make some predictions on the measured scattering intensity based upon the wavefunction in Eq. (3.1).

We first remark that the overall normalization of the wave function in Eq. (3.1) is irrelevant to the current discussion, the meaningful parameter being the relative amplitudes of the incident and scattered wavefunctions.

---

[2]In this time-independent treatment (see Appendix 3A) the indexes i and f incident and scattered wave's wavevectors, i.e. respectively the one upstream and downstream of the sample, respectively.

We now try to tackle two critical questions:

(1) The first question is how many incident particles cross a plane perpendicular to the wavevector per unit time and unit area, i.e. what irradiance the incident beam has. This number is proportional to the so-called probability flux, which can be determined using simple rules known from Quantum Mechanics. We recall here that, given a propagating wavefunction $\psi(\boldsymbol{r})$ defined in the $\boldsymbol{r}$-space, the associated current density, or the probability flux, is $\boldsymbol{J} = \rho\boldsymbol{v}$, where $\rho = |\psi(\boldsymbol{r})|^2$ and $\boldsymbol{v}$ is the propagation speed of the wave. More appropriately, we can consider the radial current as $J = \boldsymbol{J} \cdot \hat{\boldsymbol{r}}$. Clearly, this quantity depends on the normalization factor used for the wavefunction; considering the normalization in, given that incident particles are photons, one has $J = c$, where c is the speed of light in vacuum.

(2) The second question is how many scattered photons cross an infinitesimal area around the $\hat{\boldsymbol{k}}_f$-direction, which subtends a solid angle $d\Omega$. Again, this number is proportional to the related photon current density, which, based on the amplitudes defined in Eq. (3.1), is equal to:

$$J_f = c|f(\boldsymbol{k}_i, \boldsymbol{k}_f)|^2 d\Omega = J|f(\boldsymbol{k}_i, \boldsymbol{k}_f)|^2 d\Omega. \qquad (3.2)$$

It appears that the current of scattered particles flowing inside solid angle $d\Omega$ around the direction $\hat{\boldsymbol{k}}_f$ is equal to the incident photon current, J, reduced by a factor $|f(\boldsymbol{k}_i, \boldsymbol{k}_f)|^2 d\Omega$. This suggests that, among all incident photons, the ones scattered inside a fan of solid angle $d\Omega$ are those ultimately crossing the perpendicular area $d\sigma = |f(\boldsymbol{k}_i, \boldsymbol{k}_f)|^2 d\Omega$; this suggests introducing the new variable:

$$\frac{d\sigma}{d\Omega} = |f(\boldsymbol{k}_i, \boldsymbol{k}_f)|^2, \qquad (3.3)$$

which is customarily referred to as the differential cross-section of the scattering process. Therefore, if the detector ultimately covers a

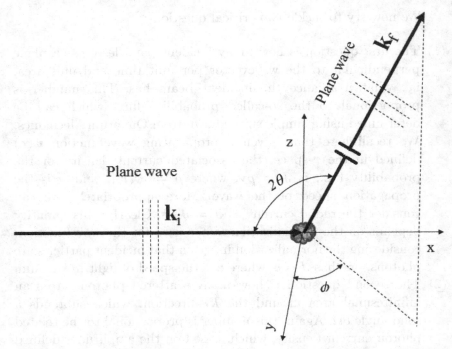

Figure 3-1:   A schematic rendering of the scattering process and the plane wave approximation (see text).

solid angle $d\Omega$, one has:

$$\frac{d\sigma}{d\Omega} = \frac{\text{Rate of photons scattered into } d\Omega}{J \cdot d\Omega}. \tag{3.4}$$

Notice that J is the rate of impinging photons per unit area in the MKS unit system, measured in photons/s/m^2.

Thus far, we made reference to the time-independent treatment of the scattering problem discussed in Appendix 3A. A more comprehensive treatment of the scattering process must reflect the dynamics of the scattering units, i.e. the movements of electrons, which, in turn, are strictly related to those of the atoms they belong to. The occurrence of an internal microscopic dynamics in the target sample gives rise to inelastic effects — i.e. to energy transfers between the particle beam and the sample. Time-dependent effects can be tackled by treating the sample-probe interaction Hamiltonian in a perturbative fashion, as detailed further below.

It is important to recognize that the much lower momentum carried by incident photons compared to target scatterers makes the $|\boldsymbol{k}_i| \approx |\boldsymbol{k}_f|$ approximation always safe, at least for all *IXS* measurements considered in this book. Nonetheless, inelastic effects do affect photons albeit they typically lead to small $(10^{-8} - 10^{-6})$ relative variation of $|\boldsymbol{k}_i|$.

When dealing with inelastic scattering phenomena, the relevant variable is the double differential scattering cross-section:

$$\frac{d^2\sigma}{d\Omega\,dE_f} = \frac{\text{Rate of photons scattered into } d\Omega \text{ with final energy between} E_f \text{ and} E_f + dE_f}{J d\Omega dE_f}, \qquad (3.5)$$

whose integral in the variable E_f yields the differential cross-section previously introduced.

In a typical scattering measurement, the detector covers a solid angle small enough to justify the assumption that the scattered photons impinging on its sensitive area are parallel, i.e. they have the same wavevector $\boldsymbol{k}_f$, which is also orthogonal to the wavefront. In summary, the scattering event probed by the measurement entails a transition of the photon states between two distinct plane waves, as schematically illustrated in Figure 3-1.

In principle, the scattering intensity could be thus computed within a perturbative approach, while counting all scattering plane waves emanating from the same incident plane wave and having wavevector $\boldsymbol{k}_f$ pointing to a direction 2θ within the covered solid angle, $\Delta\Omega$. However, this strategy poses a problem, as it requires a proper normalization of the involved wave functions, but, unfortunately, plane waves have normalization integral that diverges for long distances.

To circumvent this difficulty, it is useful to represent the scattering problem in a cubic box of size L while using periodic boundary conditions, and, eventually, consider the limit of large L. This expedient also makes the counting of the $\boldsymbol{k}_f$ states more straightforward, as discussed further below. Given the confinement of the problem within a cubic box with periodic boundary conditions, a normalized plane reads as $\psi(\mathbf{r}) = (1/\sqrt{L^3})\exp(i\mathbf{k}\cdot\mathbf{r})$.

3.3. The crucial role of the cross-section in *IXS* measurements (Scopigno *et al.*, 2005)

Let us now try to quantitatively predict the number of scattered photons detected by an *IXS* measurement. We assume here that such a measurement is carried out in transmission geometry, as it is often the case of real *IXS* experiments. For the sake of simplicity, we will further assume that: (1) the sample has the shape of a slab of thickness t_s, (2) the beam impinges orthogonally to it, and (3) its orthogonal section, Σ_B, is constant across the sample. The latter assumption amounts in discarding the finite divergence (focusing) of the incident beam.

Finally, we will assume that the beam only illuminates a portion of the cross-sectional area of the sample, which, again, is a fairly typical scenario for *IXS* measurements. It is essential to recognize that the beam intensity is not constant across the sample thickness, as a part of it gets absorbed by the sample itself. This intensity reduction can be evaluated by considering the scattering intensity from an infinitesimal sample slice orthogonal to the beam, having constant thickness dx, parallel to the sample front surface and located (downstream) at a distance x from it. The rate of photons scattered by such an infinitesimal sample volume, which have an energy in the bandwidth ΔE_f, and directions in a solid angle $\Delta\Omega$ reads as:

$$dN_f\left(x\right) = \dot{N}_i n_s \Sigma_B dx \ \exp(-\mu x)\left(\frac{d^2\sigma}{d\Omega\, dE_f}\right)$$

$$\times \Delta\Omega\Delta E_f \ \exp[-\mu(t_s - x)] \tag{3.6}$$

where n_s is the number of scattering units for unit volume and μ the absorption coefficient. Notice that in the above equation we have considered that the scattering from the infinitesimal sample slab is absorbed both downstream and upstream of it. Overall, the scattering contribution from the whole sample can be simply obtained by integrating the above equation in the variable x between

0 and t_s, which yields:

$$\dot{N}_f(x) = \dot{N}_i n_s \Sigma_B t_s \exp(-\mu t_s) \left(\frac{d^2\sigma}{d\Omega\, dE_f} \right) \Delta\Omega\, \Delta E_f. \qquad (3.7)$$

As mentioned, this formula is strictly derived for a forward scattering geometry; however, it can be easily extended to the case of finite scattering angle by using simple trigonometry.

For a given *IXS* experiment, the optimal length of the sample is the one maximizing the scattering intensity, as identified by the condition $\partial \dot{N}_f / \partial t_s = 0$; the latter yields $t_s = 1/\mu$, which shows that the optimal sample thickness should match the absorption length, $1/\mu$, of the sample. For typical incident beam energies of current *IXS* spectrometers, and for samples having atomic number $Z > 4$, the attenuation of the incident intensity is primarily due to the photoelectric absorption process. The photoabsorption length typically decreases upon increasing Z, and this implies that *IXS* measurements on low Z materials require the use of relatively large samples, with a typical thickness in the *cm* range. The achievement of an optimal sample thickness might become prohibitive for samples available in a small amount or that must be embedded in small volumes, as is typically the case of high-pressure experiments in Diamond Anvil Cells (*DACs*).

Given this preliminary discussion, we now focus on the analytical derivation of the *IXS* double differential cross-section $d^2\sigma/d\Omega\, dE_f$, which is the only factor in Eq. (3.7), which conveys insight on the dynamic response of the target sample.

3.4. From the Hamiltonian of the scattering process to the double differential cross section (Sinha, 2001, Scopigno *et al.*, 2005)

The description of the interaction between the photon beam and the target sample can be approached starting from a suitable expression of the vector potential $\mathbf{A}(\mathbf{r})$ of the electromagnetic field associated with the impinging photon beam. This variable can be written as a linear combination of normalized plane waves, its explicit form being

(Sinha, 2001):

$$A(r) = \sqrt{\left(\frac{2\pi\hbar}{\omega_k L^3}\right)} \sum_{k,\alpha} c\hat{\varepsilon}_\alpha \left[a_{k,\alpha} \exp(i\boldsymbol{k} \cdot \boldsymbol{r}) + a^\dagger_{k,\alpha} \exp(-i\boldsymbol{k} \cdot \boldsymbol{r})\right].$$

$$(3.8)$$

Here the indexes 'k' and 'α' label, respectively, the wave vector and polarization of the plane wave representing the photon state, while the operator $a_{k,\alpha}$ and its Hermitian conjugate, $a^\dagger_{k,\alpha}$ are, respectively, the annihilation and creation operators acting on the photon state (k, α); the propagation speed of the photon and its angular frequency, are c (speed of light in vacuum) and ω_k, respectively. Notice that the positive and negative arguments of the exponentials in Eq. (3.8) respectively define either an upstream or a downstream propagating plane wave.

Once $A(r)$ has been defined, we can now focus on the Hamiltonian describing the interaction between incident radiation and the electrons belonging to the sample's atoms, here portrayed as points charges e having mass m_e. Neglecting relativistic effects and the coupling between photons and electron spin, such a Hamiltonian reads as (Sinha, 2001):

$$H = H_{\text{el}} + H^{(1)}_{\text{int}} + H^{(2)}_{\text{int}}, \qquad (3.9)$$

in which the unperturbed term reads as

$$H_{\text{el}} = \sum_j \left[\frac{\boldsymbol{p}_j^2}{2m_e} + W(\boldsymbol{r}_i)\right] + W^{e-e}_{\text{int}}. \qquad (3.10)$$

Here, $\boldsymbol{r}_j$ and $\boldsymbol{p}_j$ are the position and the momentum of the ith electron, respectively, $W(\boldsymbol{r}_j)$ is the potential energy of the jth electron, and W^{e-e}_{int} is the electron-electron interaction potential integrated over the electron clouds of the target atoms. The two terms accounting for the perturbation induced by the impinging electromagnetic radiation are:

$$H^{(1)}_{\text{int}} = -\frac{e}{2m_e c} \sum_j \left\{A(\boldsymbol{r}_j), \boldsymbol{p}_j\right\} \qquad (3.11)$$

and

$$H_{\text{int}}^{(2)} = \frac{1}{2} r_0 \sum_j \boldsymbol{A}^2(\boldsymbol{r}_j). \tag{3.12}$$

Here the symbol $\{,\}$ denotes the anticommutator operator, while $r_0 = e^2/(m_e c^2)$ is the classical electron radius in cgs units.

As defined in Eq. (3.8), the vector potential $\boldsymbol{A}(\boldsymbol{r})$ is a linear combination of creation and annihilation operators. Therefore, to the leading order, the terms linear in $\boldsymbol{A}(\boldsymbol{r})$ (Eq. (3.11)), account for processes in which a photon is either annihilated (absorption) or created (emission). We can call these processes one-photon processes, as a single photon exists either only before or only after the process involved.

Conversely, perturbative terms proportional to the square of the vector potential — as the Thomson term in Eq. (3.12) — contain the products $\boldsymbol{a}_{k,\alpha} \boldsymbol{a}_{k,\alpha}^\dagger$ or $\boldsymbol{a}_{k,\alpha}^\dagger \boldsymbol{a}_{k,\alpha}$. These operator products first create and then annihilate a photon, or vice versa; in either case, leaving the number of photons unchanged. Scattering processes clearly fit in this category, as a photon exists before (incident photon) and after (scattered photon) the scattering event; processes of this kind are referred to as "two-photon" processes.

However, the terms linear in the vector potential of Eq. (3.11) describe two-photon processes to the second-order, since, to this order, they become quadratic in the vector potential.

In these second-order processes, a photon is absorbed in the transition from the initial atomic state to an intermediate one; then the transition to the final atomic state is accompanied by the emission of another photon. For incident energies far from an electron energy resonance, these second-order terms are entirely dominated by the Thomson term, which thus provides the leading contribution to the scattering signal. Hereafter, we will assume that scattering processes can be entirely described by this term alone.

Counting the photon states

As mentioned, the confinement of the scattering problem to a cubic box of finite size L makes the counting of the final photon states

k_f relatively straightforward. To compute them explicitly (Bransden and Joachain 1983), we need an explicit formula for the density of k_f-states.

Within the L-sided box and considering the periodic boundary conditions, the only allowed wavevector values are:

$$k_x = \frac{2\pi}{L}n_x, \quad k_y = \frac{2\pi}{L}n_y, \quad k_z = \frac{2\pi}{L}n_z, \tag{3.13}$$

where n_x, n_y and n_z are generic integers. Let us now consider the allowed wavevector states per plane wave polarization, within a solid angle $d\Omega$, and having energy included between E_f and $E_f + dE_f$. These states are included in a reciprocal space spherical shell between the radii k_f and $k_f + dk_f$ (see Figure 3-2), whose volume is:

$$dV(k_f) = 4\pi k_f^2 dk_f.$$

The set of wavevectors defined in Eqs. (3.13) forms a lattice in the k-space, whose periodicity is defined by a unit cell having a volume $V_{\min} = (2\pi/L)^3$. Since there is a single point in the volume $V_{\min}$ the quantity $V_{\min}^{-1} = L/(2\pi)^3$ defines the density of k-points in the reciprocal space. It follows that the number of lattice points (wavevector states) in the reciprocal space within the elemental shell is given by the ratio $dV(k_f)/V_{\min}$. Let $d^2n/d\Omega dE_f$ be the density of photon states (for each polarization state[3]), having energy between E_f and $E_f + dE_f$ and wavevector direction in the solid angle $d\Omega$. Overall, one has:

$$\frac{d^2n}{d\Omega\,dE_f}d\Omega\,dE_f = k_f^2\left(\frac{L}{2\pi}\right)^3 d\Omega\,dk_f = \frac{k_f^2}{\hbar c}\left(\frac{L}{2\pi}\right)^3 d\Omega\,dE_f,$$

where the second identity is obtained by differentiating the dispersion law $E_f = \hbar c k_f$, which leads to $dk_f = (\hbar c)^{-1}dE_f$. By comparing first

---

[3]In this treatment the polarization state of the photon is properly accounted for by the versor $\hat{\varepsilon}_\alpha$ in the vector potential of Eq. (3.8), which determines the interaction Hamiltonian and, via the Fermi Golden Rule, the double differential cross-section.

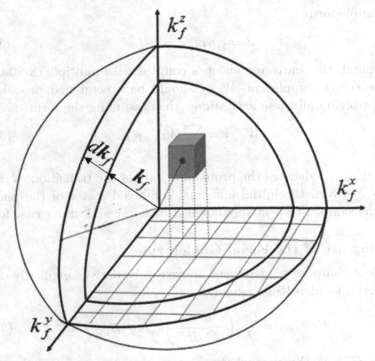

Figure 3-2:   The shell in the reciprocal space considered when counting (see text) the photon states. Here the cube enclosing a lattice point represents the unit cell of size $V_{min} = (2\pi/L)^3$ (see text).

and last members of the above identity chain, one can finally obtain:

$$\frac{d^2n}{d\Omega\,dE_f} = \frac{L^3}{8\pi^3}\frac{k_f^2}{\hbar c}. \tag{3.14}$$

The double differential cross-section can be now expressed as:

$$\frac{d^2\sigma}{d\Omega\,dE_f} = \left[\frac{1}{J}\frac{d^2n}{d\Omega\,dE_f}\right] W_{i\to f}, \tag{3.15}$$

where $W_{i\to f}$ is the rate of the transition — *i.e.* the transition probability for unit time — between initial and final states of the photon. Also, in the considered plane wave representation of the scattering problem, the amplitude of the photon current assumes the simple form:

$$J = |\psi(\mathbf{r})|^2 c = c/L^3. \tag{3.16}$$

In general, the scattering event is coupled with multiple excitations in the target sample and $W_{i \to f}$ should be a combined probability rate involving all these excitations, thus assuming the form:

$$W_{i \to f} = \sum_{I,F} W_{I,i \to F,f}. \tag{3.17}$$

Here $W_{I,i \to F,f}$ denotes the probability rate of the transition, $|I, i\rangle \to |F, f\rangle$, between the initial and final combined states of the photon and the sample, respectively labelled by lower and upper case fonts.

Making use of the Fermi Golden rule

The probability rate introduced above can be derived explicitly using the Fermi Golden Rule, namely:

$$W_{I,i \to F,f} = \frac{2\pi}{\hbar} \left(\frac{d^2 n}{d\Omega \, dE_f} \right) |\langle F, f|H_{int}|I, i\rangle|^2, \tag{3.18}$$

where H_{int} is the perturbative interaction term of the Hamiltonian, here assumed coincident with the Thomson term defined by Eq. (3.12). This term can be made explicit by inserting in it the vector potential in Eq. (3.8), while taking $\omega(k)$ equal to ck_i and ck_f, in the initial and final photon states, respectively. Furthermore, the combination of the previous five equations leads to the following expression for the double differential cross-section:

$$\frac{\partial^2 \sigma}{\partial\Omega \, \partial E_f} = r_0^2 \frac{k_f}{k_i} (\hat{\varepsilon}_i \cdot \hat{\varepsilon}_f)^2$$

$$\times \sum_{F,I} P_I \left| \langle F| \sum_j \exp(i\mathbf{Q} \cdot \mathbf{r}_j)|I\rangle \right|^2 \delta(\hbar\omega + E_I - E_F), \tag{3.19}$$

where, as mentioned, $\mathbf{r}_j$ is the position of the jth electron in the target atom. Notice that in the above formula we added the term

$\delta(\hbar\omega + E_I - E_F)$, which ensures the energy conservation in the scattering process, while the momentum conservation is embodied by the identity $\hbar Q = \hbar(k_f - k_i)$. In this treatment the energy gained by the photons is equal to $-\hbar\omega$, while the δ-function ensures that $-\hbar\omega = -(E_F - E_I)$, with $E_F - E_I$ being the energy gained by the sample.

It appears that the cross-section entails a sum over all initial states of the system, where the factor P_I defines the statistical population of these initial states. This average coincides with the ordinary thermal average of the system at equilibrium.

3.5. The cross-section in the adiabatic approximation (Bransden and Joachain, 1983, Sinha, 2001, Scopigno *et al.*, 2005)

The right-hand side of Eq. (3.19) contains three independent factors: (1) the square of the electron radius, r_0^2; (2) the coefficient $(k_f/k_i)(\hat{\varepsilon}_i \cdot \hat{\varepsilon}_f)^2$, which is determined by the plane wave parameters defining the photon states before and after the scattering event, and (3) the integral term, which directly relates to the sample properties.

In general, the latter can hardly be handled analytically, due to the complex interplay between electronic and nuclear coordinates; however, it becomes easily treatable on an analytical ground under the following approximations:

• The centre of mass of the electronic cloud drifts with no delay in response to the slow nuclear motion. This assumption, customarily referred to as adiabatic or Born-Oppenheimer approximation (Bransden and Joachain, 1983), authorizes the factorization of the target system "ket" state as $|S\rangle = |S_n\rangle|S_e\rangle$, where nuclear and electronic states are labelled by the suffixes "n" and "e" respectively. The accuracy of this assumption ultimately owes to the difference between nuclear and electronic masses and the correspondingly different timescales defining their dynamics. Also, this approximation is valid when the energy exchange is smaller than excitation energies of electrons in bound core states, which includes all cases of practical interest for this book. In liquid metals, this condition is only violated for electron densities near the Fermi level.

- As $|S_e\rangle$ is unaffected by the scattering process, the difference between the initial $|I\rangle = |I_n\rangle|I_e\rangle$ and the final $|F\rangle = |F_n\rangle|F_e\rangle$ states of the sample atoms is uniquely due to excitations associated with density fluctuations involving atomic nuclei.

Within the validity of these assumptions, the double differential cross-section reduces to:

$$\frac{d^2\sigma}{d\Omega\,dE_f} = r_0^2\frac{k_f}{k_i}(\hat{\varepsilon}_i \cdot \hat{\varepsilon}_f)^2 \sum_{F_n,I_n} P_{I_n} \left| \langle F_n| \sum_j f_j(\boldsymbol{Q}) \exp(i\boldsymbol{Q}\cdot\boldsymbol{R}_j)|I_n\rangle \right|^2$$
$$\times \delta(\hbar\omega + E_{I_n} - E_{F_n}) \tag{3.20}$$

where E_{I_n} and E_{F_n} are the energies associated with the initial and final nuclear states, respectively, and

$$f_j(\boldsymbol{Q}) = \langle F_e \left| \sum_{\alpha=1}^{Z} \exp(i\boldsymbol{Q}\cdot\boldsymbol{r}_j^\alpha) \right| I_e\rangle \tag{3.21}$$

is the atomic form factor of the jth atom, whose center of mass is located at $\boldsymbol{R}_j$. Here $\boldsymbol{r}_j^\alpha$ is the space coordinate of the αth electron in the centre-of-mass frame of the jth atom, while the symbol $|I_e\rangle$ labels the ground state of the electronic wave function of a given atomic nucleus. Considering the delocalized nature of the electron, the form factor of the jth atom can also be expressed as a volume integral over the electron clouds:

$$f_j(\boldsymbol{Q}) = \int d\boldsymbol{r}\, \exp(i\boldsymbol{Q}\cdot\boldsymbol{r})\rho_j(\boldsymbol{r}), \tag{3.22}$$

which shows that $f_j(\boldsymbol{Q})$ is the Fourier transform of the electronic cloud of the jth atomic species. Since we are considering monatomic systems, all atoms have the same form factor, i.e. $f_j(\boldsymbol{Q}) \equiv f(Q)$, which further simplifies the expression of the double differential cross-section. The latter now reduces to:

$$\frac{\partial^2 \sigma}{\partial\Omega\,\partial E_f} = K \sum_{F_n, I_n} P_{I_n} \left| \langle F_n | \sum_j \exp(i\boldsymbol{Q}\cdot\boldsymbol{R}_j)|I_n\rangle \right|^2$$
$$\times\, \delta(\hbar\omega + E_{F_n} - E_{I_n}) \tag{3.23}$$

where $K = r_0^2 (k_f/k_i (\hat{\varepsilon}_i \cdot \hat{\varepsilon}_f)_f^2 |f(Q)|^2$.

Here we also assume that, with the system being isotropic, the form factor uniquely depends on $Q = |\boldsymbol{Q}|$.

It is usually safe to approximate $f(Q)$ with the value calculated for a free atom, i.e. in the perfect gas phase. In fact, the leading contribution to $f(Q)$ mainly arises from core electrons which are more tightly bound to the atomic nucleus, and their orbits are insensitive to the aggregation phase of the sample. For $Q \to 0$, $f(Q)$ reaches its maximum value $f(0) = \int d\boldsymbol{r}\rho_j(\boldsymbol{r}) = Z$; as Q increases, it sharply decreases, thus causing a substantial reduction of the *IXS* scattering intensity.

The above expression of the cross-section consists of three independent factors involving either atomic or nuclear coordinates and can be cast in a more compact form after the few additional manipulations briefly illustrated in Appendix 3B. The expression eventually achieved,

$$\frac{d^2\sigma}{d\Omega\,dE_f} = N \frac{r_0^2}{\hbar} \left(\frac{k_f}{k_i}\right) (\hat{\varepsilon}_i \cdot \hat{\varepsilon}_f)^2 |f(Q)|^2 S(Q,\omega), \tag{3.24}$$

contains the dynamic structure factor:

$$S(Q,\omega) = \frac{1}{2\pi N} \int_{-\infty}^{\infty} dt \sum_{j,k=1}^{N} \langle \exp\{i\boldsymbol{Q}\cdot[\boldsymbol{R}_j(t) - \boldsymbol{R}_j(0)]\}\rangle \exp(-i\omega t).$$
$$\tag{3.25}$$

Also, the above formula can be rewritten by keeping in mind the definition of the variable density in the reciprocal space (see

Eq. (2.74)), which yields:

$$S(Q,\omega) = \frac{1}{2\pi N} \int_{-\infty}^{+\infty} dt \langle \delta n(-\boldsymbol{Q},0)\delta n(\boldsymbol{Q},t) \rangle \exp(-i\omega t). \quad (3.26)$$

The identity above entails the replacement of the microscopic density $n(\boldsymbol{Q},t) = \sum_{j=1}^{N} \exp[-i\boldsymbol{Q}\cdot\boldsymbol{R}_j(t)]$ with its fluctuation from equilibrium $\delta n(\boldsymbol{Q},t) = \sum_{j=1}^{N} \exp[-i\boldsymbol{Q}\cdot\boldsymbol{R}_j(t)]-n\delta(\boldsymbol{Q})$. As discussed in Chapter 2, this replacement is justified as long as the elastic forward scattering is discarded, which is the case of all scattering measurements.

It can be readily noticed that, as expected for a double differential cross-section, the left-hand member of Eq. (3.24) is dimensionally an area/energy; in fact, all factors enclosed in brackets therein are dimensionless, while $S(Q,\omega)$ has the dimension of ω^{-1}, therefore $(r_0^2/\hbar)S(Q,\omega)$ has the dimension of area/energy.

We finally remark that the expression of the cross-section in Eq. (3.24) was strictly derived within the Born–Oppenheimer approximation for a target sample composed of N identical atoms. When different atomic species are instead present, the derivation of the scattering cross-section is similar, provided the system is isotropic, i.e. invariant for rotations. The "effective" form factor, in this case, results from the average value of the form factors of different atoms.

The general case of a system composed of molecules with a more or less pronounced anisotropy is instead less straightforward and makes the computation of the cross-section slightly more involved (see, e.g. Ref. (Egelstaff, 1967)). A more detailed treatment of this problem within the hypothesis of random molecular orientations and weak coupling between orientational and translational degrees of freedom leads to the conclusion that the spectrum splits into a coherent and an incoherent component. Namely:

$$\frac{\partial^2 \sigma}{\partial\Omega\,\partial\omega} = K\{\langle F^2(Q)\rangle_\Omega S_C(Q,\omega) + \delta\langle F^2(Q)\rangle_\Omega S_I(Q,\omega)\}, \quad (3.27)$$

where $\delta\langle F(Q)^2\rangle_\Omega = \langle F(Q)^2\rangle_\Omega - \langle F(Q)\rangle_\Omega^2$, where the suffix "$\Omega$" indicates that the average $\langle\cdots\rangle_\Omega$ is performed over molecular orientations. The suffixes "I" and "C" label the coherent and the incoherent parts of the dynamic structure factor, respectively.

3.6. An estimate of the count rate

Once an analytical expression for the cross-section has been explicitly derived, it is now useful to estimate the count rate practically achievable in an *IXS* measurement. To this purpose, one should start from expressing the total scattering cross-section associated with the whole scattering volume, while explicitly considering photo-absorption processes. The flux of scattered photons in the solid angle $\Delta\Omega$ and the energy interval ΔE_f is thus given by:

$$d\dot{N}_f = \dot{N}_i \frac{n_s}{2.71828 \cdot \mu} \frac{d^2\sigma}{d\Omega\, dE_f} \Delta\Omega\, \Delta E_f, \qquad (3.28)$$

which is obtained from Eq. (3.7), while assuming an optimal thickness $t_s = 1/\mu$, for which the exponential attenuation factor (see Eq. (3.7)) reduces to $1/e = 1/2.71828$. At this stage, both members of the equation can be integrated in time and the double integration over both solid angle and final energy must be performed to eventually obtain the total cross-section of the *IXS* scattering. Namely:

$$\frac{N_f}{N_i} = \frac{n_s}{2.71828 \cdot \mu} \int \frac{d^2\sigma}{d\Omega dE_f} d\Omega dE_f. \qquad (3.29)$$

Let us consider the low Q limit, where the atomic form factor is approximately $f(Q) \approx Z$. If one considers Eq. (3.24) with $(k_f/k_i)(\hat{\varepsilon}_i \cdot \hat{\varepsilon}_f)^2 = 1$, the integral in the left hand side of the above equation is equal to $S(0)(Zr_0)^2$. Therefore, in this limit, one has:

$$\frac{N_f}{N_i} \propto \frac{(Zr_0)^2 n_s}{\mu} = \frac{\sigma_C}{\sigma_A}, \qquad (3.30)$$

where $\sigma_C = (Zr_0)^2$, while $\sigma_A = n_s/\mu$ is the absorption cross-section. An idea of the count efficiency of *IXS* technique is provided by Figure 3-3, which displays for an incident X-Ray beam of 22 keV, the value of the ratio between the total number of photons contributing to the Thomson scattering and those lost through

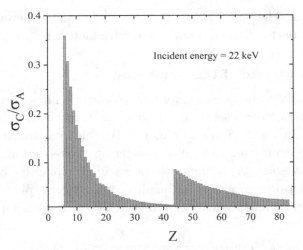

Figure 3-3:   Ratio between the Thomson cross section and the attenuation cross section inclusive of all extinction processes, including photoelectric absorption; values reported refer to a generic sample having optimal thickness; they are computed for different values of the atomic number Z and for an incident X-Ray beam of 22 keV. Data are taken from Ref. (Scopigno *et al.*, 2005).

all other processes, including photoelectric absorption. An abrupt variation is particularly evident when the incident energy reach the K-edge, i.e. the binding energy of the the innermost electron shell; indeed, core electrons are those primarily interacting with the incident X-Ray.

Typical energies at which *IXS* spectrometers are operated are close to the 22 keV considered in Figure 3-3. At these energies, and for target atoms with $Z > 4$, intensity losses of the incident beam are almost completely determined by the photoelectric absorption whose attenuation is proportional to Z^4, aside from the abrupt changes (discontinuities) when the energy approaches the electron absorption thresholds. The presence of a strong absorption drastically reduces the volume of illuminated sample for high Z materials. This volume decrease is only partially compensated by the Z^2 increase of the Thompson cross section. This makes the *IXS* measurements in low Z materials more challenging than in their low Z counterparts. All these estimates rest not only on the $Q = 0$ approximation, but also on the assumption that single scattering events yield the dominating intensity.

The intensity contribution from multiple scattering events depends on both sample and spectrometer features and its explicit estimate is not straightforward. However, the ratio between single scattering and double scattering intensities can be explicitly computed within an essentially forward scattering detection geometry, i.e., again, for $Q \approx 0$. An upper estimate of such a ratio is given by:

$$R_{1,2}(Q = 0) \leq \frac{\pi \Sigma_B n_s r_0^2}{Z^2 S(0)} \int_0^\pi d\theta' [f^2(\theta') S(\theta')] \qquad (3.31)$$

The integral on the righthand side accounts of double scattering events, in which X-Rays are first deviated by an angle $2\theta'$ within the $[0, \pi]$ interval and then travel parallel the incident beam. After the first scattering event, the static structure factor is expressed as $S(\theta') = S((4\pi/\lambda)\sin(\theta'))$; a similar expression was used for the form factor $f(\theta')$. In the upper estimate proposed above, the beam attenuation of the second scattering event is not properly considered. It appears that in a scattering measurement, the mitigation of the relative contribution of double scattering event requires in the first place a minimization of the cross-sectional of the beam Σ_B.

Appendix 3A: The scattering problem: A time-independent theoretical description

We illustrate here the theoretical treatment of the scattering of plane waves by an anisotropic medium; this treatment deals with the static case, in which no time-dependence is considered. Let us first consider the time-independent Schrödinger equation describing a quantum particle of mass m impinging on a region where a potential energy $V(r)$ exists. This potential might represent the interaction experienced by a flying electron while approaching an atom or a neutron in the proximity of an atomic nucleus. In general, we can consider $V(r)$ as a continuous source of scattering confined within a volume V, which can be regarded as the equivalent of the sample in a real scattering experiment. The time-independent Schrödinger

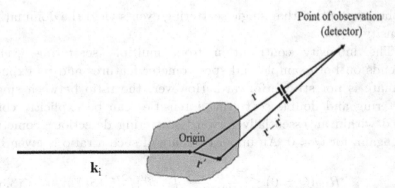

Figure 3-4:   A schematic layout of the scattering problem discussed in the text. The shadowed area defines the region of space where a limited-range potential $V(\mathbf{r})$ exists.

equation reads as:

$$\left[\frac{\hbar^2}{2m}\nabla^2 + E_k\right] u_k(\mathbf{r}) = V(\mathbf{r})u_k(\mathbf{r}). \tag{3A.1}$$

Upstream of the scattering, the plane wave is in a region where $V(\mathbf{r}) = 0$ (see the scheme in Figure 3-4) and the Schrödinger equation assumes the homogeneous form:

$$\left[\frac{\hbar^2}{2m}\nabla^2 + E_i\right] u_{\mathbf{k}i}(\mathbf{r}) = 0, \tag{3A.2}$$

whose solution is the incident plane wave $u_{\mathbf{k}i}(\mathbf{r}) = \exp(i\mathbf{k}_i \cdot \mathbf{r})$, where the wavevector $\mathbf{k}_i$ has amplitude $k_i = \sqrt{2mE_i}$.

The non-homogeneous Schrödinger equation (3A.1) can be tackled analytically using the Green function method, where the Green function $G(\mathbf{r})$ is the solution of the source point equation:

$$\left[\frac{\hbar^2}{2m}\nabla^2 + E_i\right] G(r, k_i) = \delta(\mathbf{r}). \tag{3A.3}$$

In general, the solution of the above equation is not unique since we can add to it any solution of the corresponding homogeneous equation in Eq. (3A.2), as it can be immediately verified by summing both members of the two equations. Therefore, we can conceive such a solution as a sum containing $\exp(i\mathbf{k}_i \cdot \mathbf{r})$. In particular, the integral expression:

$$u(\mathbf{r}) = \exp(i\mathbf{k}_i \cdot \mathbf{r}) + \int_V d\mathbf{r}' G(\mathbf{r} - \mathbf{r}', k_i) V(\mathbf{r}') u(\mathbf{r}'). \qquad (3A.4)$$

is also a solution to Eq. (3A.1), as it readily follows from applying the operator $[(\hbar^2/2m)\nabla^2 + E_i]$ to both members of the above equation. In the above equation, $u(\mathbf{r})$ appears both in the left-hand side and in the integral on the right-hand side and the integral is extended over the scatterer volume, whose generic point is located at the position $\mathbf{r}'$. If $V(\mathbf{r}')$ is small, such an integral equation can be solved by an iterative method. However, this strategy is not convenient since the solution is not unique, and we want to select here only those solutions making physical sense for the scattering problem we are considering.

Indeed, we should exclude the non-physical solution of the Shroedinger equation corresponding to an incoming wave, i.e. a scattering waves going toward the region where $V(\mathbf{r}) > 0$, instead of emanating from it. In order to discriminate the propagation direction of the scattering wave, we should express the Green function in terms of its Fourier $\mathbf{k}$-components. In other terms, we should deal with the Fourier transform of the Green function:

$$G(r, k_i) = \frac{1}{(2\pi)^3} \int d\mathbf{r}' \frac{\exp(i\mathbf{r}' \cdot \mathbf{r})}{[E_i - \hbar^2 k^2/(2m)]} = -\frac{m}{4\pi^3 \hbar^2} \int d\mathbf{k}' \frac{\exp(i\mathbf{k}' \cdot \mathbf{r})}{(k'^2 - k_i^2)}$$
$$(3A.5)$$

The circumstance that this Green's function fulfills Eq. 3A.3 can be easily verified by applying the operator $[(\hbar^2/2m)\nabla^2 + E_i]$ to the first integral above. This cancels the denominator in the integral, while leaving as the only surviving term $1/(2\pi)^3 \int d\mathbf{k}' \exp(i\mathbf{k}' \cdot \mathbf{r})$. The latter term indeed coincides with the $\delta(\mathbf{r})$ in the righthand side of 3A.3. Note that the Green function now depends only on the moduli of the variables $\mathbf{r}$ and $\mathbf{k}_i$, i.e. $G(\mathbf{r}, \mathbf{k}_i) \equiv G(r, k_i)$, as the integration in the righthand side of the above formula is performed over all directions. Let us now express the integral in the righthand side of the above equation in a more explicit form:

$$\int d\boldsymbol{k}' \frac{\exp(i\boldsymbol{k}' \cdot \boldsymbol{r})}{k_i^2 - k'^2}$$

$$= 2\pi \int_0^\infty k'^2 dk' \int_{-1}^1 d(\cos\theta) \frac{\exp(ik'r\cos\theta)}{k'^2 - k_i^2}$$

$$= \frac{2\pi}{ri} \left\{ \int_0^\infty k' dk' \left[\frac{\exp(ik'r)}{k'^2 - k_i^2} \right] - \int_0^\infty k' dk' \left[\frac{\exp(-ik'r)}{k'^2 - k_i^2} \right] \right\}$$

$$= \frac{2\pi}{ri} \int_{-\infty}^\infty k' dk' \left[\frac{\exp(ik'r)}{k'^2 - k_i^2} \right],$$

where the last member of this identity chain follows from the change of variable $k' \rightarrow -k'$. The insertion of this result into Eq. (3A.5) leads to:

$$G(r, k_i) = -\frac{m}{2\pi^2 \hbar^2 ri} \int_{-\infty}^\infty k' dk' \frac{\exp(ik'r)}{k'^2 - k_i^2}. \qquad (3A.6)$$

The integral above has the evident problem of diverging for $k' = \pm k_i$. To have a definite integral, we need to bypass these poles, and this can be done by writing the integral as follows:

$$G_\pm(r, k_i) = \lim_{\varepsilon \to 0} \left[-\frac{m}{2\pi^2 \hbar^2 ri} \int_{-\infty}^\infty k' dk' \frac{\exp(ik'r)}{k'^2 - (k_i^2 \pm i\varepsilon)} \right]. \qquad (3A.7)$$

If $\varepsilon > 0$, the function under integral has no poles in the real axis but only above and below in the plane of a complex variable z which extends k' analytically in the complex plane, i.e. such that $Re z = k'$. These poles can be easily identified using the identity:

$$z^2 - k_i^2 \pm i\varepsilon = \left(z - \sqrt{k_i^2 \pm i\varepsilon} \right) \left(z - \sqrt{k_i^2 \pm i\varepsilon} \right).$$

Since we are considering the limit $\varepsilon \to 0$, then the poles are located at:

$$\sqrt{k_i^2 \pm i\varepsilon} \approx \pm \left[k_i^2 \pm \frac{i\varepsilon}{2k_i} \right].$$

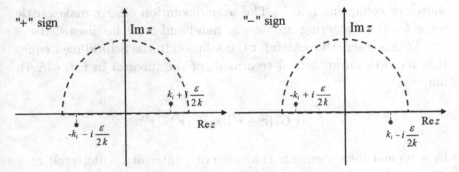

Figure 3-5: A scheme of the contours corresponding to the "−" and "+" options of the $\pm i\varepsilon$ term in the Green function in Eq. (2.44) (see text).

At this stage, we can evaluate the integral in Eq. (3A.9) following a semicircular arc closed by a straight path overlapping with the real axis, as shown in Figure 3-5 and making use of the method of residues.

We have chosen a semicircles belonging to the upper half-plane $\Im z > 0$ as the integral along this path is vanishing for diverging radius, while it would have been diverging if chosen in the $\Im z < 0$ semi-plane. Notice that the horizontal branch of such a closed contour — i.e. the one coincident with the real axis — does not intercept the poles as these were displaced by the addition of an imaginary component.

As reported in Figure 3-5, if we chose the imaginary part with the "+" sign, the first pole will be included in the $Im z < 0$ half-plane; vice-versa, if we chose the "−" sign, the second pole would be in the $Im z < 0$ half-plane. The points belonging to the $Im z < 0$ half-plane are those excluded by the integration path along the arc. In other terms, taking the limit $\varepsilon \to 0$ and using the Jordan lemma (Hildebrand, 1962), one eventually obtains:

$$G_{\pm}(r, k_i) = -\frac{m}{2\pi\hbar^2} \exp(\pm i k_i r), \qquad (3A.8)$$

where, as discussed, the solutions with the sign "+" and "−" respectively correspond to the $+i\varepsilon$ and $-i\varepsilon$ options in Eq. (3A.7), which describe a spherical wave either emitted from the scattering source or collapsing into it. The second solution clearly makes little sense for the scattering problem at hand and must be discarded.

At this stage, the search for the solution to the Schrödinger equation requires an analytical treatment of the integral in Eq. (3A.4), namely:

$$\int_V d\mathbf{r}' G(|\mathbf{r} - \mathbf{r}'|, k_i) V(\mathbf{r}') u(\mathbf{r}').$$

In a typical measurement, the scattering intensity is detected at a distance from the sample much larger than the sample size, and the $\mathbf{r} \gg \mathbf{r}'$ condition is valid. Within the validity of this approximation, the Green function,

$$G(|\mathbf{r} - \mathbf{r}'|, k_i) = -\left(\frac{m}{2\pi\hbar^2}\right)\frac{\exp(-ik_i|\mathbf{r} - \mathbf{r}'|)}{|\mathbf{r} - \mathbf{r}'|}, \qquad (3A.9)$$

is simplified by taking $|\mathbf{r} - \mathbf{r}'| \approx |\mathbf{r}| = r$ in the denominator. Unfortunately, the same approximation cannot be adopted in the numerator. This owes to the significant variations of $\mathbf{r}'$, which cause rapid changes of the phase, defined by the imaginary exponent. Phase changes with $\mathbf{r}'$ are critical as they have a dramatic impact on the outcome of the integral in Eq. (3A.11). For this reason, it is important that the contribution of the integral at $\mathbf{r}'$ is taken with the correct phase.

To treat the exponent in the numerator, we can instead consider the following expansion:

$$|\mathbf{r} - \mathbf{r}'| = \sqrt{r^2 - 2rr'\cos\theta + r'^2}$$

$$= r\sqrt{1 - 2\frac{r'}{r}\cos\theta + \frac{r'^2}{r^2}} \approx r - \mathbf{r}\cdot\hat{\mathbf{r}}', \qquad (3A.10)$$

where θ is the angle between $\mathbf{r}$ and $\hat{\mathbf{r}}'$. Hence, one can write:

$$k|\mathbf{r} - \mathbf{r}'| = k_i r - \mathbf{k}_f \cdot \hat{\mathbf{r}}',$$

where $\mathbf{k}_i = \mathbf{k}$ and $\mathbf{k}_f = k_i\hat{\mathbf{r}}$ are the wavevectors upstream and downstream of (and very far from) the scattering event, respectively.

Notice that the amplitude of the wavevector upstream and downstream of the scattering is unchanged, i.e. $|\mathbf{k}_i| = |\mathbf{k}_f|$. This is not surprising as, in this description, time-dependent (inelastic) effects are absent, and the target sample is assumed infinitely massive and at rest, with its constituents being frozen in their initial position. As mentioned, for the *IXS* measurements considered in this book, inelastic effects, although crucial, have a weak relative strength; therefore the approximation $|\mathbf{k}_i| \approx |\mathbf{k}_f|$ is accurate. In summary, the Green function assumes the following form:

$$G(|\mathbf{r} - \mathbf{r}'|, k_i) \approx G(r, k_i) = \frac{m}{2\pi\hbar^2} \frac{\exp(-ik_i r)}{r} \exp(i\mathbf{k}_f \cdot \mathbf{r}').$$

Within this so-called far-field approximation, the profile of the wave function in Eq. (3A.4) thus reads as:

$$u(\mathbf{r}) = \exp(i\mathbf{k}_i \cdot \mathbf{r}) - \frac{m}{2\pi\hbar^2} \left[\frac{\exp(-ik_i r)}{r}\right]$$
$$\times \int_V d\mathbf{r}' \exp(i\mathbf{k}_f \cdot \mathbf{r}') u(\mathbf{r}') V(\mathbf{r}'), \qquad (3A.11)$$

which can be cast in the more compact form:

$$u(\mathbf{r}) = \exp(i\mathbf{k} \cdot \mathbf{r}) + f(\mathbf{k}_i, \mathbf{k}_f) \frac{\exp(ik_i r)}{r}, \qquad (3A.12)$$

where

$$f(\mathbf{k}_i, \mathbf{k}_f) = -\frac{m}{2\pi\hbar^2} \int_V d\mathbf{r}' \exp(-i\mathbf{k}_f \cdot \mathbf{r}') u(\mathbf{r}') V(\mathbf{r}')$$

has the dimension of a length and is customarily the scattering length.

The Born approximation

The above equation can be handled analytically with an iterative approach. The first step is customarily referred to as the Born approximation, and it corresponds to replacing $u(\mathbf{r})$ in the integral with its zero order expression, $u(\boldsymbol{r}) = \exp(i\boldsymbol{k}_i \cdot \boldsymbol{r})$, which yields:

$$u(\boldsymbol{r}) = \exp(i\boldsymbol{k}_i \cdot \boldsymbol{r}) - \frac{m}{2\pi\hbar^2}\left[\frac{\exp(ik_i r)}{r}\right]$$

$$\int_V d\boldsymbol{r}'\exp(i\boldsymbol{k}_i \cdot \boldsymbol{r}')\exp(-i\boldsymbol{k}_f \cdot \boldsymbol{r}')V(\boldsymbol{r}').$$

The above equation can be written in the following, more compact form:

$$u(\boldsymbol{r}) = \exp(i\boldsymbol{k}_i \cdot \boldsymbol{r}) + \frac{\exp(ik_i r)}{r}f_B(\boldsymbol{k}_i, \boldsymbol{k}_f), \qquad (3\text{A}.13)$$

where $f_B(\boldsymbol{k}_i, \boldsymbol{k}_f) = -m/(2\pi\hbar^2)V(\boldsymbol{Q})$ and $V(\boldsymbol{Q}) = \int_V d\boldsymbol{r}' \exp(i\boldsymbol{Q} \cdot \boldsymbol{r}')V(\boldsymbol{r}')$ with $\boldsymbol{Q} = \boldsymbol{k}_i - \boldsymbol{k}_f$. Notice that $V(\boldsymbol{Q})$ is simply the Fourier transform of the interaction potential $V(\boldsymbol{r})$.

Appendix 3B: A compact expression for the double differential cross-section in the adiabatic approximation

Eq. (3.23) provides an expression of the double differential cross-section which we report here for convenience:

$$\frac{d^2\sigma}{d\Omega\, dE_f} = K \sum_{F_n, I_n} P_{I_n}\left|\langle F_n|\sum_j \exp(i\boldsymbol{Q} \cdot \boldsymbol{R}_j)|I_n\rangle\right|^2$$

$$\times \delta(\hbar\omega + E_{I_n} - E_{F_n}), \qquad (3\text{B}.1)$$

where $K = r_0^2\left(\frac{k_f}{k_i}\right)(\hat{\varepsilon}_i \cdot \hat{\varepsilon}_f)^2|f(Q)|^2$.

In the following, we illustrate the various manipulations of this formula which ultimately lead to the more compact expression of the cross-section of Eq. (3.24).

(1) Integral representation of the δ-function of energy

First, one can use the integral definition of the δ-function and the general property $\delta(\hbar\omega) = \delta(\omega)/\hbar$. This identity enables one to rewrite the δ-function term, ensuring the conservation of energy, as:

$$\delta(\hbar\omega + E_{I_n} - E_{F_n}) = 1/2\pi\hbar \int_{-\infty}^{\infty} dt\exp\{-i[\omega + (E_{I_n} - E_{F_n})/\hbar]t\} .$$

Hence:

$$\frac{d^2\sigma}{d\Omega\,dE_f} = K \sum_{F_n,I_n} P_{I_n} \left| \langle F_n | \sum_{j=1}^{N} \exp(i\boldsymbol{Q}\cdot\boldsymbol{R}_j)|I_n\rangle \right|^2$$

$$\delta(\hbar\omega + E_{I_n} - E_{F_n})$$

$$= \frac{K}{2\pi\hbar} \sum_{F_n,I_n} P_{I_n} \sum_{j=1}^{N} \left[\int_{-\infty}^{\infty} dt \langle I_n | \exp(-i\boldsymbol{Q}\cdot\boldsymbol{R}_k)|F_n\rangle \right.$$

$$\times \langle F_n | \exp[i(E_{F_n}/\hbar)t] \exp(i\boldsymbol{Q}\cdot\boldsymbol{R}_j)$$

$$\left. \exp[-i(E_{I_n}/\hbar)t]|I_n\rangle \right] \exp(-i\omega t).$$

(2) Heisenberg representation of a time-dependent operator

Given that $|I_n\rangle$ and $|F_n\rangle$ are the eigenstates of the Hamiltonian H corresponding to the respective eigenvalues E_{I_n} and E_{F_n}, one can write $\exp(-iE_{I_n}t/\hbar)|I_n\rangle = \exp(-iHt/\hbar)|I_n\rangle$ and $\langle F_n|\exp(iE_{F_n}t/\hbar) = \langle F_n|\exp(iHt/\hbar)$.

Hence:

$$\frac{d^2\sigma}{d\Omega\,dE_f} = \frac{K}{2\pi\hbar} \sum_{F_n,I_n} P_{I_n} \left| \langle F_n | \sum_j \exp(i\boldsymbol{Q}\cdot\boldsymbol{R}_j)|I_n\rangle \right|^2$$

$$\times \delta(\hbar\omega + E_{I_n} - E_{F_n})$$

$$= \frac{K}{2\pi\hbar} \sum_{F_n,I_n} P_{I_n} \sum_{jk=1}^{N} \int_{-\infty}^{\infty} \langle I_n | \exp(-i\boldsymbol{Q}\cdot\boldsymbol{R}_k)|F_n\rangle$$

$$\times \left\langle F_n \left| \exp\left(i\frac{H}{\hbar}t\right) \exp(i\boldsymbol{Q}\cdot\boldsymbol{R}_j) \exp\left(-i\frac{H}{\hbar}t\right) \right| I_n \right\rangle$$

$$\times \exp(-i\omega t).$$

Notice that here $\boldsymbol{R}_j = \boldsymbol{R}_j(0)$ is the position of the jth atomic nucleus at the time $t = 0$. At this stage, it is useful to consider the Heisenberg representation of the time evolution of a variable $A(t)$, namely $A(t) = \exp(iHt/\hbar)A(0)\exp(-iHt/\hbar)$. This enables us to write:

$$\exp\left[i\mathbf{Q}\cdot\mathbf{R}_j(t)\right] = \exp\left[i\left(H/\hbar\right)t\right]\exp\left(i\mathbf{Q}\cdot\mathbf{R}_j\right)\exp\left[-i\left(H/\hbar\right)t\right].$$

Combining the above manipulations, one has:

$$\frac{d^2\sigma}{d\Omega\,dE_f} = \frac{K}{2\pi\hbar} \times \sum_{F_n,I_n} P_{I_n} \sum_{jk=1}^{N} \int_{-\infty}^{\infty}$$

$$\langle I_n|\exp[-i\mathbf{Q}\cdot\mathbf{R}_k(0)]|F_n\rangle$$

$$\langle F_n|\exp[i\mathbf{Q}\cdot\mathbf{R}_j]|I_n\rangle\exp(-i\omega t).$$

(3) Completeness of the final eigenstate

At this stage, the equation above can be further developed by considering that the final eigenstates of the system form a complete set, *i.e.* $\sum_{F_n}|F_n\rangle\langle F_n| = I$, with I being the identity operator. This expedient makes trivial the sum over the final, after-scattering, states. If we consider the (k,j)-th term of the double sum above:

$$\frac{d^2\sigma}{d\Omega\,dE_f} = K/(2\pi\hbar)\sum_{F_n,I_n} P_{I_n}\sum_{jk=1}^{N}\langle I_n|\exp[-i\mathbf{Q}\cdot\mathbf{R}_k]|F_n\rangle$$

$$\times\langle F_n|\exp[i\mathbf{Q}\cdot\mathbf{R}_j(t)]|I_n\rangle$$

$$= K/(2\pi\hbar)\sum_{I_n} P_{I_n}\sum_{jk=1}^{N}\langle I_n|\exp[-i\mathbf{Q}\cdot\mathbf{R}_k]$$

$$\exp[i\mathbf{Q}\cdot\mathbf{R}_j(t)]|I_n\rangle$$

(4) The thermal average

It is finally necessary to recognize that the above expression involves a sum on both initial and final states of the system. In a scattering measurement, these states are unknown since only the final and initial states of the photon can be measured. However one knows that both energy and momentum of the probed sample system must be conserved in the scattering process and that the sample is a many-atoms system at thermal equilibrium, instead of being prepared in a well-defined quantum state. In this respect, it is appropriate to replace the weighted average $\sum_{I_n} P_{I_n} ...$ by an ordinary "thermal" average, which leads to the following identification:

$$\frac{d^2\sigma}{d\Omega \, dE_f} = (K/2\pi\hbar) \sum_{I_n} P_{I_n} \langle I_n | \exp\left[-i\boldsymbol{Q} \cdot \boldsymbol{R}_k\right] \exp\left[i\boldsymbol{Q} \cdot \boldsymbol{R}_j(t)\right] | I_n \rangle$$

$$= (K/2\pi\hbar) \sum_{jk=1}^{N} \langle \exp\left[-i\boldsymbol{Q} \cdot \boldsymbol{R}_k\right] \exp\left[i\boldsymbol{Q} \cdot \boldsymbol{R}_j(t)\right] \rangle,$$

where the last equality stems from the definition of the equilibrium correlation function of the two operators A and B as $\langle A(0)B(t) \rangle = N \sum_{I_n} P_{I_n} \langle I_n | A(0) B(t) | I_n \rangle$.

In summary, after the manipulations described above (items 1–4), considering that operators involved are commuting (classical case), the double differential cross-section in Eq. (3B.1) eventually reduces to:

$$\frac{\partial^2\sigma}{\partial\Omega \, \partial E_f} = \frac{K}{2\pi\hbar} \int_{-\infty}^{\infty} dt \sum_{jk=1}^{N} \langle \exp\{i\boldsymbol{Q} \cdot [\boldsymbol{R}_j(t) - \boldsymbol{R}_k]\} \rangle \exp(-i\omega t).$$

Using the definition of the dynamic structure factor $S(Q, \omega)$, the cross-section can be cast in a more compact form of Eq. (3.24), which is reported below:

$$\frac{d^2\sigma}{d\Omega \, dE_f} = \left(\frac{r_0^2}{\hbar}\right) \left(\frac{k_f}{k_i}\right) (\hat{\varepsilon}_i \cdot \hat{\varepsilon}_f)^2 |f(Q)|^2 N S(Q, \omega).$$

References

Bransden, B. H. & Joachain, C. J. 1983. Physics of Atoms and Molecules, London ; New York, Longman.

Egelstaff, P. A. 1967. An Introduction to the Liquid State, London, New York, Academic P.

Fowler. 2003. Notes on Scattering Theory [Online]. Available: http://galileo. phys.virginia.edu/classes/752.mf1i.spring03/ScatteringTheory.htm.

Jackson, J. D. 1999. Classical Electrodynamics, 3rd Ed, New York, Wiley.

Mobilio, S., Boscherini, F. & Meneghini, C. 2016. Synchrotron Radiation, Springer.

Sakurai, J. J. & Commins, E. D. 1995. Modern Quantum Mechanics, Revised Edition. AAPT.

Scopigno, T., Ruocco, G. & Sette, F. 2005. *Rev. Mod. Phys.*, **77**, 881.

Sinha, S. K. 2001. *J. Phys. Condens. Matter*, **13**, 7511.

Chapter 4

Complementary aspects of *IXS* and *INS*

4.1. Generalities on the *INS* technique (Lovesey, 1984, Squires, 2012)

Inelastic Neutron Scattering (*INS*) is a spectroscopic method spanning a dynamic domain which strongly overlaps with the *IXS* one. Both methods ultimately enable a direct determination of the spectrum of density fluctuations, but have different and complementary strengths, as discussed further below in this chapter.

Although an analytical derivation of the *INS* cross-section is a topic beyond the scope of this book, which is illustrated in excellent monographs on the subject (Lovesey, 1984, Squires, 2012), we will use a few relevant results of such treatment for comparison with the case of *IXS* already discussed in the previous chapter.

One of the main advantages of using neutrons as a probe is their lack of charge, which makes the interaction of neutrons with their target — the atomic nuclei — active only for short distances and enhances their penetration depth inside matter. For this reason, neutrons represent an ideal and nondestructive probe of the bulk properties of condensed matter systems.

In a neutron scattering experiment, a beam of neutrons is sent to the target sample with well-defined energy; these neutrons are schematizable as freely flying particles whose (kinetic) energy E_n is simply related to the velocity $\mathbf{v}_n$ through:

$$E_n = \frac{m_n v_n^2}{2}, \tag{4.1}$$

83

where m_n is the neutron mass; more generally, the suffix n will be used to label neutron-related quantities throughout this introductory section.

Also, the neutron beam is suitably collimated, i.e. the impinging neutrons have nearly parallel velocities. Apart from the small initial divergence of their trajectories and the one possibly acquired upon focusing, the incident neutrons can be assimilated to plane waves having wavevector $\boldsymbol{k}_n$ and carrying a momentum $\boldsymbol{p}_n = \hbar \boldsymbol{k}_n$. Also, the wavevector relates to the energy or the velocity of the neutron through $k_n = \sqrt{2m_n E_n/\hbar^2} = m_n v_n/\hbar$. Equivalently one has:

$$E_n = \frac{\hbar k_n^2}{2m_n}, \tag{4.2}$$

and likewise, the energy determines the neutron wavelength through $\lambda_n = \hbar/\sqrt{2m_n E_n}$.

In neutron research facilities, the source can be either pulsed, in which case it generates discrete bunches of freely flying neutrons via a spallation process, or continuous as in the case of neutron reactors. In the latter case, the source continuously generates neutrons by fission and shoots them all over the solid angle at very high energy. After emission, the neutrons pass through a medium containing many light atoms (moderator), which, after many collisions, substantially reduces their velocity. These collisions eventually lead the emitted neutrons to "thermalize" with the microscopic constituents of the moderator.

Upon thermalization, the kinetic energy of the emitted neutrons matches the mean kinetic energy of moderator molecules, primarily determined by the temperature of the moderator. For instance, if the latter is at ambient temperature, the moderated neutrons will eventually attain a kinetic energy of about 25 meV, which is comparable to the typical energies of phonon excitations in solids and collective excitations in liquids. Furthermore, the wavelength of thermal neutrons is a fraction of a nanometer, i.e. comparable with interatomic spacing in condensed matter.

Eventually, a suitably shaped (monochromatized and collimated) neutron beam is transported from the source to the spectrometers by steering devices as, for instance, neutron guides.

As previously done for *IXS*, the observable accessed by an *INS* measurement can be derived by handling the double differential cross-section analytically. This treatment (Lovesey, 1984) eventually reveals the direct link of this variable with the spectrum of density fluctuations, which, as discussed, is also the variable accessed by *IXS*.

Although the observable of the two spectroscopic probes are equivalent for monatomic samples, the situation is more complicated when dealing with molecular ones; in this case, the partial scattering intensities from the various atomic species in the target molecules contribute differently to the total *INS* and *IXS* signals, and this makes the comparison between *IXS* and *INS* measurements less straightforward.

The wave-particle duality of neutrons lends itself to an impressive variety of monochromatization and energy analysis options. For instance, since the energy of the neutron beam depends on the velocity of the neutrons, a possible monochromatization strategy is to select their velocities. This can be implemented, e.g., by a system of rotating slits (choppers) whose openings are suitably phased to accept only neutrons travelling with the desired speed. This monochromatisation method exploits the corpuscular nature of neutrons, here conveniently portrayed as flying bullets.

On the other hand, the energy of the incident neutron beam can be alternatively selected by Bragg reflections from crystal surfaces, as usually done for X-Rays, that is using monochromatization schemes which exploit the wave nature of neutrons.

An in-depth discussion of neutron monochromators and energy analyzers goes well beyond the scope of this book, and the interested reader can refer, for instance, to Refs. (Willis and Carlile, 2017) and (Windsor, 1981).

4.2. The cross-section of inelastic neutron scattering (see (Lovesey, 1984))

When deriving an explicit expression for the neutron scattering cross-section, the first concern is to identify a suitable form for the Hamiltonian describing the interaction between the target sample and the neutron beam impinging on it. In principle, this endeavour would require a rigorous and comprehensive theory of the nucleon-nucleon interaction, which unfortunately is not available in the literature. However, it is well-assessed experimentally that this interaction is very short-ranged and isotropic and, as such, it can be described in principle by a single unknown parameter: the nuclear scattering length b. This parameter varies in a non-trivial fashion, not only with the atomic number, but also with the nuclear mass of the scatterers. This dependence is one of the main differences between *INS* and *IXS*, since, as discussed in the previous chapter, the *IXS* cross-section does not depend directly on the atomic (molecular) mass of the scattering unit. In other terms, different isotopic species of the same material have identical *IXS* scattering cross-sections and different *INS* cross-sections. The neutron scattering cross-section also depends on the relative orientation of neutron and nuclear spins, differing in this aspect from its *IXS* counterpart.

If we assume, for the sake of simplicity, that the particles involved in the scattering event are spinless, the only analytical model of the neutron-nucleus interaction yielding, in Born approximation (see Appendix 3A), a perfectly isotropic scattering is a δ-function of space. In particular, a suitable candidate to describe such an interaction is the so-called Fermi pseudopotential:

$$V(\mathbf{r}) = \frac{2\pi\hbar^2}{m_n} \sum_{i=1}^{N} b_i \delta[\mathbf{r} - \mathbf{R}_i(t)], \qquad (4.3)$$

where $\mathbf{R}_i(t)$ is the coordinate of the ith target nucleus.

The above equation defines an extremely irregular interaction field, which diverges at the position of the ith atom nucleus and vanishes elsewhere, as to be intuitively expected when considering the collision-like interaction between point masses.

Similar to what is discussed in Chapter 3, the cross-section can be ultimately built by casting the Fermi Golden Rule in explicit form, i.e. including in it the appropriate density of final states and using, as an interaction matrix element the potential computed between initial and final states, as labelled by the respective wavevector states $\mathbf{k}_i$

and k_f, which reads as:

$$\langle k_f | V(r) | k_i \rangle = \frac{2\pi\hbar^2}{m_n} \sum_i b_i \exp(-i Q \cdot R_i(t)). \qquad (4.4)$$

As explicitly shown in Ref. (Lovesey, 1984), the following expression for the double differential *INS* cross-section can be eventually derived:

$$\left. \frac{\partial^2 \sigma}{\partial\Omega\partial\, dE_f} \right|_{INS} = \left(\frac{k_f}{k_i} \right) \frac{N}{\hbar} \left\{ \overline{b}^2 S_C(Q,\omega) + \delta(b^2) S_I(Q,\omega) \right\}, \qquad (4.5)$$

where $\delta(b^2) = \left[\overline{b^2} - \overline{b}^2 \right]$ is the mean square fluctuation of the scattering length b of the target system, where the overbar symbol labels the average over the atoms of the sample. Such a fluctuation arises from the possible coupling of neutron and nuclear spin (if not vanishing) and to the possibly mixed isotopic composition of the sample, if it is multi-isotopic.

Generalities on the comparison between *IXS* and *INS* techniques

Overall, *IXS* and *INS* methods present several analogies, as follows:

(1) They both probe bulk properties of materials at variance with spectroscopic methods using strongly interacting particles such as electrons, which primarily convey insight on surface properties of the samples.
(2) More specifically, they probe the dynamics of density fluctuations via the Fourier transform of their autocorrelation function, i.e. the dynamic structure factor, $S(Q,\omega)$.
(3) They are "mesoscopic" spectroscopies, i.e. they map distances and timescales matching, respectively, separations between first adjacent atoms (molecules) and cage oscillation periods.

Other, more formal, similarities and differences emerge from the comparison of the *IXS* double differential cross-section in Eq. (3.25) and its *INS* counterpart in Eq. (4.5), which, for instance, shows that:

- Both *INS* and *IXS* cross-sections contain the k_f/k_i ratio as a prefactor. However, as discussed in the following, for *IXS*, $k_f \approx k_i$ and this factor can be thus safely discarded in the frequency window spanned by typical inelastic measurements.
- In many aspects, the role of the form factor $f(Q)$ for the *IXS* cross-section mirrors that of the scattering length b for the *INS* one. However, these properties respectively define either the strength of the photon-electron electromagnetic interaction, or the one of neutron-nuclei interaction. As a consequence, $f(Q)$ — ultimately determining the coupling strength between X-Ray photon and target matter — increases upon increasing the number of electrons Z; yet, it doesn't depend on the mass of target atoms. Therefore, the *IXS* cross-section, at odds with the *INS* one, is insensitive of the isotopic species of the target sample. Furthermore, $f(Q)$ sharply decreases upon increasing Q, until it vanishes, typically within a few tens of nm^{-1}. This stems from the circumstance that the volume of the electronic cloud coupling with incident photon decreases as Q^{-3}. Conversely, b remains substantially constant until Q becomes comparable with the inverse of the nuclear diameter. Consequently, the *INS* cross-section substantially decreases only at Q values much larger than its X-Ray scattering counterpart.

Other significant differences will be discussed in the following sections; in particular, the next one will be focused on the substantially different effects that the so-called kinematic constraints have on *IXS* and *INS* measurements.

4.3. Kinematic limitations (Squires, 2012)

In principle, the dynamic response of a sample can be probed as an average over different length- and time-scale domains by proper tuning of the variables Q and ω. However, the values accessible for these variables are subjected to some limitations arising from the energy and momentum conservation in the scattering event. These limitations are usually referred to as "kinematic" as they stem from the superimposition of conservation laws ruling the kinematic of the sample-probe collision. As shown below, these limitations affect neutron and X-Ray scattering to a very different extent. Let's discuss the case of neutrons first.

(a) Neutron scattering

The momentum conservation in the neutron-nucleus scattering event imposes the following constraint on the exchanged wavevector:

$$Q = k_i - k_f. \tag{4.6}$$

Here, as in Chapter 3, the suffix "i" and "f" label the value of the variable before and after the scattering event, respectively. The equation above can be manipulated by squaring both members, which yields:

$$Q^2 = k_f^2 + k_i^2 - 2k_i k_f cos(2\theta)$$

where 2θ is the scattering angle already introduced in Chapter 3. Both members of the above equation can be now divided by k_i^2, thus obtaining:

$$(Q/k_i)^2 = (k_f/k_i)^2 + 1 - 2(k_f/k_i) cos(2\theta). \tag{4.7}$$

At this stage, one can use the energy conservation law:

$$\hbar\omega = E_i - E_f,$$

and the circumstance that a neutron travelling with velocity $v_{i,f} = \hbar k_{i,f}/m_n$ has a (kinetic) energy:

$$E_{i,f} = \hbar^2 k_{i,f}^2/(2m_n). \tag{4.8}$$

The combination of the last two equations yields:

$$E_{i,f} = \hbar^2 k_{i,f}^2/(2m_n). \tag{4.9}$$

Notice that the square root on the right-hand side is real-valued under the condition $\hbar\omega \leq E_i$, which corresponds to the obvious constraint that the energy transferred from the neutron to the sample cannot exceed the initial neutron energy. Furthermore, by combining this equation with Eq. (4.7), one has:

$$Q = k_i \sqrt{2 - \hbar\omega/E_i - 2\cos(2\theta)\sqrt{1 - \hbar\omega/E_i}}. \qquad (4.10)$$

The requirement that the Q value defined by the equation above is a real number introduces further restrictions to the explored dynamic (Q, ω)-range, as represented by the curves in the bottom plot of Figure 4-1.

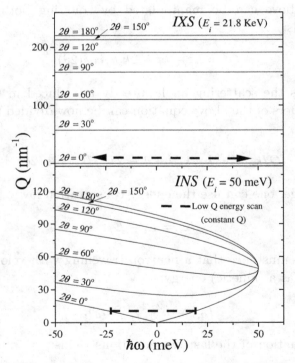

Figure 4-1:   Boundaries of the accessible dynamic range as evaluated for *IXS* and *INS* for different values of 2θ (see text). The horizontal dashed lines represent constant Q-energy scan performed at the low Q values where collective modes are dominant. It clearly appears that kinematic restrictions are relevant to *INS* measurements only.

(b) Inelastic X-Ray Scattering

Let's now discuss how kinematic limitations impact *IXS*. While, for neutrons, the dependence of energy on the wavevector is quadratic (see Eq. (4.2)), for X-Ray, it is linear; namely:

$$E_{i,f} = \hbar c k_{i,f}. \tag{4.11}$$

From the dispersion law above, it follows that the energy conservation in the scattering event can be cast in the following form:

$$k_f/k_i = 1 - \hbar\omega/E_i. \tag{4.12}$$

Combining the above equation with Eq. (4.11) and superimposing the momentum conservation (Eq. (4.6)) one obtains eventually:

$$Q = k_i\sqrt{1 + (1 - \hbar\omega/E_i)^2 - 2\cos(2\theta)(1 - \hbar\omega/E_i)}. \tag{4.13}$$

In principle, as previously observed for *INS*, the portion of the dynamic plane accessible by *IXS* is subjected to the constraint that the argument under the square root is positive, and, consequently, Q is real.

However, in practice, this does not represent a real limitation, as, for *IXS* measurements, $\hbar\omega/E_i$ values typically span the $10^{-7} - 10^{-5}$ window, thus being much smaller than the other terms under the square root. Neglecting this term is thus safe in typical *IXS* applications; the use of this approximation in the equation above yields:

$$Q = k_i\sqrt{2[1 - \cos(2\theta)]} = 2k_i\sin(\theta)$$
$$= \frac{4\pi}{\lambda_i}\sin(\theta), \tag{4.14}$$

where we used $k_i = 2\pi/\lambda_i$, with λ_i being the wavelength of the incident beam. The formula above shows that, for *IXS*, Q and ω are uncoupled and, specifically, Q depends on the scattering angle 2θ and the incident wavelength only. Consequently, there are no inherent limitations to the (Q,ω) plane virtually explorable by *IXS*, except those imposed by the resolution, which establishes a minimum finite energy exchange, $\hbar\omega$.

Of course, practical limitations still exist; for instance, for *IXS* instruments operated through temperature scans of the active optics, the accessible exchanged energy window is limited to the maximum heat load tolerated to prevent distortions of crystal planes.

As mentioned, the kinematic constraints, which stem from Eqs. (4.10) and (4.13) for *INS* and *IXS* respectively, give rise to boundary curves which delimit the accessible (Q, ω) domain to an extent which depends on the scattering angle. Examples of these boundaries are displayed in Figure 4-1. The curves reported are obtained by varying the scattering angle parametrically and are plotted in a restricted energy range comparable to that typically spanned by *IXS* and *INS* investigations on disordered systems. It appears that, at low Q values, kinematic boundaries substantially restrict the energy (frequency) domain spanned by constant-Q scans of *INS* spectra, conversely, they impose no practical limitation to the energy range typically spanned by *IXS* scans.

4.4. The roles of instrumental resolution and spectral contrast

Figure 4-2 displays a typical spectral shape measured by *IXS* on a water sample at $Q = 3$ nm^{-1}. Even discarding other spurious intensity effects, an *IXS* spectrometer, like any other spectroscopy tool, does not measure the spectrum inherently scattered by the sample, often quoted as the real or true sample spectrum, but rather its convolution with the energy resolution function (see also Section 2.6). The latter can be defined as the spectral shape detected by the instrument when the true spectrum from the sample is perfectly elastic, i.e. when it has the form of a δ-function of energy (or frequency). The resolution profile gives a measure of the instrumental contribution to the broadening of the measured spectral shape. In some respects, it can be thought of as a blurring effect in the rendering of the real sample spectrum. Consistently, the width of the resolution spectrum gives an estimate of the minimum energy separation between two discernible spectral features of comparable intensity.

In Figure 4-2, the comparison between the *IXS* signal with the corresponding resolution profile emphasizes the presence, in the

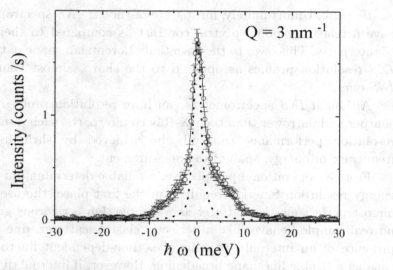

Figure 4-2: An example of how a typical *IXS* lineshape (open circles) compares with the respective resolution function (dotted line). A best fitting lineshape is also reported for reference as a continuous line. The measurement was taken at the indicated Q value on a sample of liquid water at room temperature (Cunsolo *et al.*, 1999).

water spectrum, of large spectral wings, which relate to acoustic-like modes propagating at terahertz frequencies, as discussed in successive chapters in further detail.

If two adjacent spectral features under scrutiny have very different intensities, it is the spectral contrast which determines the instrumental capability to resolve them.

The spectral contrast can be defined as the ratio between the intensities of the resolution profile at its maximum, $R(\omega = 0)$, and at a given frequency shift $R(\omega = \omega^*)$. This ratio gives a measure of the sharpness of the resolution profile by assessing "how rapidly" its wings decay upon increasing ω.

To provide an example, the proper detection of a spectral feature at ω^* whose intensity is 100 times smaller than the one of the elastic peak would greatly benefit from a spectral contrast $R(0)/R(\omega^*)$, at least as high as 100.

It appears that a high contrast is required to accurately discern weak low-frequency features gathering on the side of a dominating elastic peak. Unfortunately, most state-of-the-art *IXS* spectrometers have a relatively poor spectral contrast as compared to their *INS* counterparts. This owes to the essentially Lorentzian shape of typical *IXS* resolution profiles as opposed to the sharp, almost Gaussian, *INS* ones.

Although *INS* spectrometers can have resolution profiles much sharper and narrower than their X-Ray counterparts, such a superior resolution performance can only be achieved by shrinking the frequency, or energy, span of the measurement.

From an operational perspective, a reliable determination of the energy resolution function requires, in the first place, the use of an almost perfect elastic scatterer as a sample. On a rigorous ground, no real sample behaves like a perfectly elastic scatterer, due to the presence of an internal dynamics, i.e. time-dependent fluctuations causing a sizable lineshape broadening. However, if internal dynamic processes are sufficiently slow or if they provide a minor contribution to the spectral shape, a δ-function of frequency centred at $\omega = 0$ (perfectly elastic scattering), may give a reasonable approximation of the spectrum from the sample, which thus becomes a suitable candidate for resolution measurements.

As schematically illustrated in Figure 4-3, the spectrometer image of this prevalently elastic spectrum provides a reliable measurement of the instrumental resolution function. Of course, the need for a relatively fast and statistically accurate estimate of the resolution profile calls for the use of a strong scatterer.

These requirements often suggest measuring the resolution via the scattering from a Plexiglas sample at $Q = Q_m$ ($\approx 10\,\text{nm}^{-1}$ for Plexiglas), the position of the main $S(Q)$ maximum, where the elastic scattering intensity becomes dominant, especially for an amorphous solid.

For *INS* experiments, the resolution function is often determined as the spectrum of a vanadium standard, whose cross-section is theoretically known. This also offers the opportunity for calibration factor needed for an absolute intensity measurement of the scattering profile from the sample.

Let us now consider a spectroscopic measurement aimed at detecting inelastic modes in the spectrum of a liquid. As exemplified

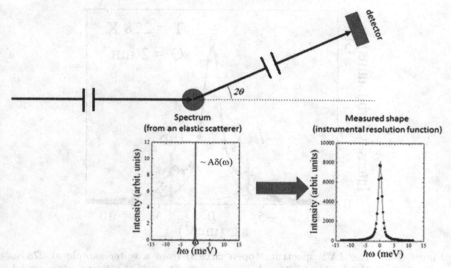

Figure 4-3: A schematic representation of the concept of instrumental energy resolutions. The reported resolution profile (squares on the left plot) is reported along with its best fitting model shape (solid line); it measured at the ID16 *IXS* beamline of the European Synchrotron Radiation Facility on a sample of Plexiglas at the Q-position of its first $S(Q)$ maximum.

by Figure 4-4, these modes often show up in the spectral shape as a pair of peaks symmetrically shifted from the elastic energy $\hbar\omega = 0$ by an amount $\pm\hbar\omega_s$. Obviously, the minimal requirement for a successful measurement is that the spanned frequency interval fully includes those excitations. If these have an acoustic origin, at sufficiently low Q's, their positions ω_s are linked to the adiabatic sound velocity c_s through $\omega_s = c_s Q$. In this perspective, the knowledge of c_s might provide a rough prediction of the chance of success of the measurement as well as some indication about the optimal instrumental set-up. Sometimes its actual value might be known or at least roughly estimated from literature measurements or thermodynamic databases.[1]

---

[1]The adiabatic sound velocity can be indeed derived from thermodynamic properties such as the isothermal compressibility, density and the specific heats ratio γ.

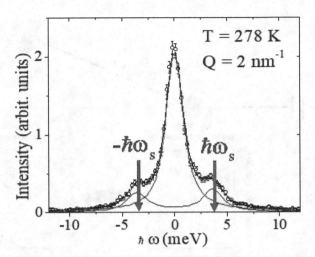

Figure 4-4:   The *IXS* spectrum (open circles) from a water sample at 278 K, measured at the indicated Q value, is compared with the best-fitting model lineshape, and its individual spectral components. The plot emphasizes the presence of a collective mode having a dominant frequency $\pm\hbar\omega_s$.

If c_s is too high, kinematic limitations might prevent the adequate coverage of the inelastic acoustic modes in some Q regions. Conversely, if c_s is too small, these excitations might not be resolved adequately, especially at the lowest Q's. These experimental challenges might be successfully addressed or mitigated by suitable tailoring of the instrumental setup. For instance, using a higher or lower incident beam energy or accessing to higher or lower Q's; in any case, the spectrometer configuration, and the corresponding dynamic domain explorable, are the result of a trade-off between competing — "resolution" and "kinematic" — limitations. However, as discussed, the latter limitations are essentially irrelevant for *IXS* measurements, while imposing severe penalties to *INS* ones. Still, for *IXS* as well, practical challenges limit the extension of the Q-range reachable, both at low and high Q's. For instance, the high-Q decay of the form factor poses major challenges to *IXS* measurements exploring the so-called single particle limit (see Chapter 7). Furthermore, if frequency scans are implemented through the rocking or heating of crystals, the spanned angular or temperature range are usually limited by the need of preserving a superior resolution performance.

4.5. Three-axis and time of flight techniques (Windsor, 1981, Squires, 2012, Shirane *et al.*, 2002)

As opposed to constant-2θ frequency-scans of *IXS* intensity, similar scans of *INS* intensity, in general, are not performed at constant Q. In fact, according to Eq. (4.10), an ω-change directly translates into a Q-change and this $Q - \omega$ coupling usually represents an inconvenience, as most lineshape models apply to constant-Q cuts of $S(Q, \omega)$.

A possible way to circumvent this obstacle is the operation of neutron three-axis spectrometers (*TAS*) in a constant-Q mode. In this kind of measurements, 2θ and ω are scanned in a coordinated fashion, which, according to Eq. (4.10), keeps Q fixed.

Three-axis neutron spectrometry is nowadays a mature technique initially developed in the 1950s by Brockhouse (Brockhouse and Stewart, 1955). The working principle of *TAS* spectrometers is relatively straightforward and hinges on three principal rotation axes: the first is located at the monochromator crystal, and rotations around it select the energy of incident photons. The second axis is at the sample position, and rotations around this axis define the momentum transfer. Finally, the third axis is at the centre of the analyzer crystal, and rotations around it identify the energy of scattered neutrons, $\hbar\omega_f$.

In a *TAS* spectrometer, initial and final wavevector values are determined by the monochromator and the analyzer axes respectively; their 2D-movements in the horizontal plane involve four degrees of freedom, while the determination of a single (Q, ω)-point involves three parameters only: the two in-plane coordinates of the vector Q and the frequency, ω. Consequently, there are infinite possible pathways to perform a constant Q scan of the scattering intensity. A more detailed description of the advantages of various *TAS* operation modes can be found, for instance in Ref. (Shirane, Shapiro *et al.*, 2002).

An alternative strategy to perform *INS* measurements makes use of time of flight (*ToF*) techniques (Windsor, 1981), which enable the

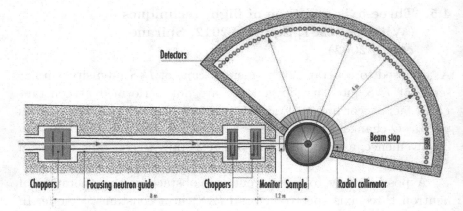

Figure 4-5: Example of an *INS* time of flight spectrometer: IN5 at the Institut Laue–Langevin. Image taken from the website: https://www.ill.eu/users/instruments/instruments-list/in5/description/instrument-layout/

fast sampling of points sparse in the (Q, ω)-plane. This technique was the first ever used to perform energy-resolved neutron scattering measurements (Dunning *et al.*, 1935), and it substantially improved thanks to the advent of reactors and pulsed sources.

In *ToF* neutron spectrometers, the energy of neutrons is determined by a selection of their velocity, often via the tuning of the angular velocities and mutual dephasing of rotating slits (choppers), or rotating crystals. These devices also set the initial time of the time-sensitive detection system, which ultimately measures the neutrons *ToF* to the detector. An example of the layout of a *ToF* spectrometer is provided by Figure 4-5.

After the scattering event, the energies of neutrons change — and so do their velocities — which causes a spread of their arrival times at the detector. A simple link exists between this time of arrival and the energy and momentum exchanged with the sample; the knowledge of such a link enables a relatively straightforward conversion of time-resolved neutron counting to an actual $S(Q, \omega)$ measurement.

Pulsed neutron sources are ideal tools to implement *ToF* methods. In this case, the pulse generation event may conveniently provide the clock start for the *ToF* analysis. Overall, *TAS* and *ToF* neutron scattering techniques offer complementary advantages:

- Neutron *TAS* techniques are particularly efficient in performing "surgical" constant Q, or constant ω, cuts of $S(Q,\omega)$ surfaces. However, these measurements are time-consuming, as each frequency-scan can probe a single Q value only. Therefore, *TAS* spectrometers are ideally suited to perform repetitive measurements of the spectrum at fixed Q while changing sample parameters such as its thermodynamic conditions, the relative concentration of a dopant or a solute, and so forth.
- Conversely, *ToF* measurements are ideal for achieving instantaneous panoramic snapshots of $S(Q,\omega)$. However, they cannot be used to perform directly constant-Q measurements of the spectrum, as required for a straightforward comparison with theoretical models of the lineshape. Of course, constant Q cuts can be generated by interpolating measured $S(Q,\omega)$ surfaces; however, this endeavour might provide uncertain results near the edge of the allowed kinematic region or if the sampling of these surfaces is too sparse.

Overall, *IXS* spectrometers represent a variant of the three-axis concept described in the previous paragraphs and schematically represented in Figure 4-6.

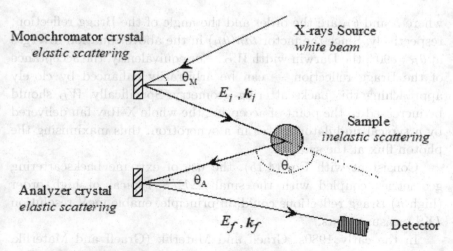

Figure 4-6:   The general scheme of the ID28 *IXS* spectrometer at ESRF.

4.6. A few critical features of *IXS* spectrometers

Most current high-resolution *IXS* spectrometers are operated with incident beam energy slightly higher than 20 keV, while collective modes in non-crystalline solids typically range in the meV window, thus they have energies smaller by about 7 orders of magnitude. Consequently, *IXS* measurements of these excitations require relative energy resolution $\Delta E/E$ at least as small as 10^{-7}. Performing *IXS* measurements with such $\Delta E/E$ values is technically challenging, and this held back for decades the development of high-resolution *IXS*, whose outset lags the one of *INS* by more than four decades.

In the late 1970s, Ando and collaborators (Ando *et al.*, 1978) demonstrated that such an outstanding performance could be achieved by using silicon crystals. To obtain the desired energy resolution while minimising photon flux penalties, the whole divergence of the incoming photon beam must be kept within the angular acceptance of the considered monochromator reflection, defined by the width of the so-called Darwin curve, W_D, of the monochromator crystal. According to the theory of X-Ray diffraction (Zachariasen, 1967), the latter reads as:

$$W_D = h(\Delta E/E)\tan(\theta_B), \qquad (4.15)$$

where h and θ_B are the order and the angle of the Bragg reflection, respectively. Since the factor $tan(\theta_B)$ in the above equation diverges at $\theta_B = 90°$, the Darwin width W_D — or equivalently, the acceptance of the Bragg reflection — can be arbitrarily enhanced by closely approaching this backscattering geometry. Specifically, W_D should be increased to the point of accepting the whole X-Ray fan delivered by a typical undulator source in a syncrotron, thus maximising the photon flux at the sample.

Consistent with Eq. (4.15), the use of extreme backscattering geometries coupled with the small $\Delta E/E$ typical of high order (high h) Bragg reflections could, in principle, enable meV resolution *IXS* measurements.

In the early 1980s, Graeff and Materlik (Graeff and Materlik, 1982) demonstrated the capability of an X-Ray backscattering scheme to reach a 10^{-7} relative resolution. This pioneering design was further developed and adapted to *IXS* applications by Dorner and collaborators (Dorner, Burkel *et al.*, 1986). These early feasibility tests of high-resolution *IXS* could at best achieve a resolving power $\Delta E/E = 5 \times 10^{-7}$ and a photon flux at the sample of about 10^6 photons/s. A slightly better performance ($\Delta E/E = 3.5 \times 10^{-7}$) was obtained with a "four bounces" crystal scheme initially developed for applications in resonant nuclear scattering with synchrotron radiation (Faigel *et al.*, 1987).

Toward the mid 1990s, the unprecedented brilliance of third-generation synchrotron radiation sources paved the way to the construction of meV-resolution *IXS* spectrometers, enabling the execution of spectral investigations on disordered systems on a routine basis.

The first meV-resolution *IXS* spectrometer ever operative was the ID16 beamline (Masciovecchio *et al.*, 1996a, Verbeni *et al.*, 1996) of the European Synchrotron Radiation Facility (ESRF) in Grenoble, France. Its transition to operations in 1995 was followed a few years later by the one of another *IXS* spectrometer with a similar design: ID28 beamline (Krisch, 2003), which nowadays is the only meV-resolution spectrometer currently active at ESRF.

Similar *IXS* spectrometers are currently available in few synchrotron sources worldwide: the **HERIX** spectrometer at the Advanced Photon Source of Argonne National Laboratory at Chicago, USA (Alatas *et al.*, 2011, Sinn *et al.*, 2001, Toellner *et al.*, 2011, see also the website https://www.aps.anl.gov/Sector-30/HERIX); the BL35U (website: http://bl35www.spring8.or.jp/) and, more recently, BL43LXU beamlines of the Spring-8 in Hiogo, Japan (Baron *et al.*, 2001; see also the website: https://user.spring8.or.jp/sp8info/?p=3138). More recently, a novel concept *IXS* beamline became operative at the National Synchrotron Light Source, NSLS-II at Brookhaven National Laboratory at Upton, NY, USA (Cai *et al.*, 2013).

In principle, the most straightforward route to perform energy (frequency) scans with an *IXS* spectrometer would be varying the Bragg angle of the analyzer or the monochromator crystal. However, any small deviation from the required extreme backscattering would rapidly degrade the energy resolution ultimately attainable. The usual strategy adopted to circumvent this problem is to scan instead the temperature of the monochromator or analyzer crystals.

In fact, for given Bragg angle and incident wavelength, the energy of a Bragg reflection uniquely depends on the d-spacing of the reflecting crystal. Consequently, varying the d-spacing of one of the two energy-selecting units (monochromator and energy analyzer), while keeping constant the other, ultimately changes the energy, $\hbar\omega$, gained/lost by the X-Ray photons after the scattering event. From the energy conservation, this energy must match the one lost/gained by the target sample in the scattering process. Therefore, scanning the temperature difference of the monochromator and analyzer units ultimately provides a measurement of the scattering signal from the sample (see Section 2.6), and more specifically, its $S(Q,\omega)$ (see Chapter 3).

The achievement of a $10^{-7} - 10^{-8}$ relative energy resolution demands millikelvin stability of the monochromator and analyzer optics, which can be achieved by using carefully designed temperature baths controlled by active feedback.

Notice that, in principle, *IXS* spectrometers suffer from the same inherent limitation of *TAS*, as they can only provide 'surgical" (constant-Q) cuts rather than a global mapping of a $S(Q,\omega)$ measurement. However, state of the art *IXS* instruments use multiple analyzers to collect the spectrum at different Q's simultaneously; this, of course, increases the throughput of the measurement substantially, thus reducing the time typically required for extensive $S(Q,\omega)$ measurements.

4.7. An example of state-of-the-art spectrometers: ID28 beamline at ESRF

To provide an example of current *IXS* spectrometers, here we discuss the relevant features of ID28 beamline (Krisch, 2003), which is the only meV-resolution spectrometer currently in operation at European Synchrotron Radiation Facility (*ESRF*), and the oldest one worldwide.

In this spectrometer, the X-Ray source consists of three undulators delivering a beam with $\Delta E/E \approx 10^{-2}$ and a 40×15 (horizontal $\times$ vertical) μrad^2 angular divergence. Such a beam is pre-monochromatized within a relative bandwidth $\Delta E/E \approx 2 \cdot 10^{-4}$ using the $Si(1, 1, 1)$ reflection from an in-vacuum cryogenically cooled Si "channel-cut" crystal. The primary duty of this pre-monochromator is to sustain the relevant portion of the heat load generated by photoelectric absorption during energy selection, thus drastically reducing the thermal stresses of the downstream optics. This minimizes thermal deformations of crystal planes of the high-resolution monochromator crystal, which drastically deteriorate the energy resolution.

The X-Ray photon beam filtered by the pre-monochromator is then back-reflected by the high-resolution monochromator, which consists of an asymmetrically cut silicon crystal oriented along the $(1, 1, 1)$ direction and operated at an 89.98^o Bragg angle. In this configuration, the angular acceptance of the monochromator exceeds the X-Ray beam divergence, which clearly minimizes photon losses in the Bragg-reflection.

The achievement of the $\Delta E/E = 10^{-7}$ goal ultimately imposes both a superior quality of the active optics and the use of high-order Bragg reflections $Si(h, h, h)$; in typical high-resolution *IXS* measurements h is included between 9 and 13. A single or a combination of mirrors can provide a bidimensional focusing of the monochromatic beam on the sample within a focal spot which can be as narrow as 30×40 (horizontal $\times$ vertical) μm^2.

The scattering from the sample is energy-analyzed by nine spherical analyzers mounted on the free extreme of the spectrometer arm, which can be rotated in the horizontal plane to reach the desired exchanged wavevector through Eq. (4.14). These analyzer units are mutually offset by the same angular amount, which enables the simultaneous collection of *IXS* spectra at different Q's.

The achievement of a high resolution impose different requirements on the characteristics of either the monochromator or the analyzer units. The low divergence of the incident X-Ray beam imposes relatively forgiving constraints on the monochromator acceptance;

conversely, with the scattering from the sample being spread all over the solid angle, the acceptance of the analyzer must be wide enough to detect a significant scattering intensity. On the other hand, an excessively large acceptance would unavoidably degrade the Q-resolution; consequently, the optimal acceptance of the analyzer results from a delicate balance between competing count rate and momentum-resolution requirements.

Considering that *IXS* measurements at low Q — for instance, a few nm^{-1} — require a Q-resolution $\Delta Q_M \leq 0.1 - 0.5\,nm^{-1}$, a reasonable value for the angular acceptance of the analyzer crystal is a few mrad, which is much larger than the Darwin width of the Bragg reflection typically used. The only option to achieve such an acceptance is granted by the use of a focusing geometry.

The need for avoiding distortions of the d-spacing ultimately degrading the energy resolution warns against the use of elastically bent crystals suggesting instead the one of a strain-free mosaic arrangement of small crystals glued on a spherical matrix. At ID28, such a mosaic is composed by $\approx 12,000$ perfect single square silicon crystals glued on a spherical surface. This spherical analyzer is operated in a 1:1 pseudo-Rowland circle geometry with aberration low enough not to appreciably affect the ultimate energy resolution (Masciovecchio *et al.*, 1996b). In this configuration, the focus of the spherical analyzer arrays is close to the sample position where the detectors are placed.

Each analyzer is equipped with upstream motorized slits, setting the momentum resolution to the desired value, and a detector with the corresponding pinhole. The detectors are Peltier-cooled 1.5 mm thick silicon diodes with a low dark count (≈ 1 count over 30 minutes).

At ID28, the scattering angle 2θ can be increased up to 55^0 and this enables the span of Q values extending up to 100 nm^{-1} while using the $Si(11, 11, 11)$ incident beam harmonic, which has 21,774 eV energy.

Figures 4-7 and 4-8 respectively display a schematic layout of ID28 beamline and a picture of its spectrometer and sample hutch.

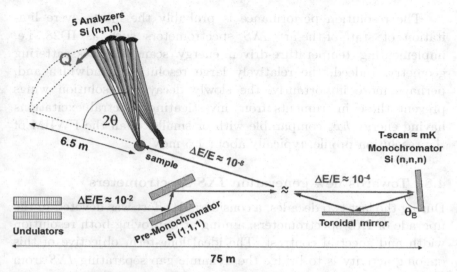

Figure 4-7: The layout of ID28 beamline *IXS* spectrometers at ESRF (see text).

Figure 4-8: A picture of the hutch containing the sample environment and spectrometer arm of the ID28 beamline at ESRF.

The resolution performance is probably the most severe limitation of state-of-the-art *IXS* spectrometers such as ID28, i.e. implementing temperature-driven energy scans in backscattering geometry. Indeed, the relatively large resolution bandwidth and, perhaps more importantly, the slowly decaying resolution wings prevent these instruments from investigating spectral excitations having energy $\hbar\omega_s$ comparable with or smaller than the FWHM of the resolution profile, typically about 1.5 meV.

4.8. Towards new generation *IXS* spectrometers

During the last two decades, a considerable effort was devoted to the upgrade of *IXS* spectrometers, aiming at improving both resolution width and spectral contrast. The ideal long-term objective of this ongoing activity is to bridge the dynamic gap separating *IXS* from low energy and momentum transfer spectroscopic techniques.

The access to such no man's land will permit the study of a large variety of still unexplored physical phenomena occurring over longer times and distances, including slow relaxation processes in highly viscous materials such as glass formers, colloids and polymers, or the collective dynamics of molecular clusters, lipid membranes, etc.

As mentioned, the most significant drawbacks of *IXS* spectrometers like the one illustrated in the previous section are a resolution profile typically broader than 1.4 meV and, perhaps more importantly, a relatively poor spectral contrast. The development of next-generation *IXS* instruments implementing a novel concept for both monochromatization and energy analysis might efficiently tackle these problems.

The use of extreme backscattering geometries and temperature-driven energy analysis makes the achievement of a ≈ 1 meV resolution challenging for two reasons: (1) it imposes demanding requirements on the temperature stability of the active optics and (2) it requires the use of incident beams delivered by high-order Bragg reflections, which penetrate deeper beneath crystal surfaces of the active optical elements, thus being more severely affected by extinction losses.

In the last decade, two optical designs exploiting the energy-dispersive character of Bragg back-reflection from asymmetrically cut

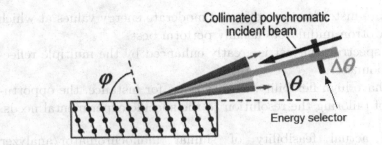

Figure 4-9: A schematic of the monochromatization principle based upon the dispersive power of crystals cut with a large asymmetry angle φ (see text). Slightly readapted from Ref. (Shvyd'ko, 2004).

crystals (Shvyd'ko, 2004) attracted increasing attention as possible alternative schemes.

Figure 4-9 illustrates the monochromatization principle implemented by using the θ-reflection from crystals having very high asymmetry, for which the reflecting surfaces form large angles with planes. The core idea is that close to backscattering, the Bragg reflection from asymmetrically cut highly angle-dispersive, as the fan of visible light exiting from an optical prism. The use of a wavelength (angle) selector, enables to filter a small slice out of the whole reflected fan, which can have, in principle, very narrow energy bandwidth.

The first is an array of four asymmetrically cut Si crystals, customarily referred to as 4-bounce monochromator (Yabashi *et al.*, 2001); the second scheme is an optical assembly including a collimator (C) a dispersive (D) and a wavelength selector (W) crystals, usually referred to as CDW or CDDW if two dispersive crystals are used instead (Shvyd'ko *et al.*, 2006).

Both optical schemes were shown to provide sub-meV wide and extremely sharp, essentially Gaussian, resolution profiles. Furthermore, these optical arrays have a wealth of advantages as compared to state-of-the-art *IXS* backscattering spectrometers, as:

- they do not require the use of high order Bragg reflections, which significantly improves their overall efficiency.

- they are instead operated at the moderate energy values at which synchrotron undulators usually perform best.
- their spectral contrast is greatly enhanced by the multiple reflection bounces.
- they have high flexibility, as they offer, for instance, the opportunity of tailoring the resolution to the specific experimental needs.

The actual feasibility of similar monochromator/analyzer schemes was first demonstrated to work for *IXS* applications in Ref. (Shvyd'ko *et al.*, 2014).

Most importantly, a novel beamline designed to achieve a meV broad and mostly Gaussian resolution function has become operative at the new synchrotron source NSLS-II at Brookhaven National Laboratory, in Upton, NY (Cai *et al.*, 2013).

The improvement of the energy resolution is also critical to access low Q values; in fact, collective modes in the spectrum can be accurately resolved provided the resolution width ΔE is narrower than their frequency $\hbar c_s Q$, which, at low Q's, becomes very small. In this perspective, there is limited use in accessing Q's lower than the threshold $Q_{min} > \Delta E / \hbar c_s$.

4.9. A closer comparison between *IXS* and *INS*

When planning a scattering experiment, the first relevant question concerns, of course, its technical feasibility. The answer often involves elaborate count rate estimates, based upon some knowledge of the actual cross-section and experimental conditions and, when possible, the reference to available literature on experiments or computational results. This feasibility assessment requires careful consideration of both resolution and kinematic limitations of the instrument, as discussed in the previous paragraphs. Even more critical is the choice of the experimental techniques best suited to the scientific scope of the measurement. While making the final choice, it is useful to keep in mind the complementary advantages and disadvantages of available experimental techniques. In the following, we will focus on the two spectroscopic methods of direct interest for this book, i.e. *IXS* and *INS*.

4.9.1. *Advantages of IXS*

Photon flux on the sample

Although tight requirements on both energy and Q resolution make *IXS* a "photon hungry" technique, state of the art *IXS* spectrometers have incident fluxes much higher than their *INS* counterparts. Precisely, they can deliver an incident photons' rate of $10^9 - 10^{10}$ photons/s within a focal spot as narrow as 100 μm^2. The rates typically attainable by inelastic neutron spectrometers are much lower ($\approx 10^5$ neutrons/s) and the focal spot much broader (often a few cm^2). Although the extremely narrow $\Delta E/E$ required by high (meV) resolution *IXS* measurements ultimately imposes severe penalties to the count rates, these remain still considerably higher than those of *INS*.

Multiple scattering

The leading contribution to the *IXS* signal attenuation is provided by photoelectric absorption, which strongly affects the count rate, especially when high Z materials are considered as samples. On the bright side, this attenuation causes a drastic suppression of multiple scattering events thus enhancing the intensity contribution from single-scattering events. This scattering signal is the one of interest for *IXS* measurements, as it is proportional to $S(Q, \omega)$. Conversely, in *INS* measurements, the scattering itself dominates the attenuation mechanism in low absorbing materials and multiple scattering events are more frequent, often yielding a significant contribution to the inelastic signal.

Transverse beam size

The incident beam of conventional *IXS* spectrometers has an extremely narrow focal spot — typically few hundreds of μm^2; this feature, combined with the strong absorption of X-Ray probes, enables measurements on smaller samples, which is a valuable asset when samples are available in small volumes only or when extreme thermodynamic conditions need to be explored. Conversely, the neutron beam is broader and more penetrating in the material's interior, making the use of tiny samples and the consequent access to extreme thermodynamic conditions very challenging for *INS*. As discussed in Section 3.6, the reduced cross-sectional area of the beam is also a valuable asset to reduce the relative intensity contribution from double-scattering events.

Beam divergence

The small divergence of *IXS* beams is a crucial prerequisite to achieve a superior Q-resolution. In a typical *IXS* measurement, such a Q-resolution depends primarily on the angular divergence of the radiation fan intercepted by the scattering collection system. Such a fan can be efficiently narrowed down by a slit, which however would also reduce the usually limited count rate. Again, it appears that the actual setting of the instrument is a delicate balance between competing resolution and count rate requirements. Finding a wise compromise is left to the experimentalist's judgement and, of course, strongly depends on the specific experimental needs.

Incoherent scattering

As discussed above, the incoherent part of *IXS* intensity mostly arises from mean square fluctuations of the form factor $f(Q)$, while for *INS* it originates from those of the scattering length b. The latter depends on the mass of the target particle, and the total (electron + nucleus and neutron) magnetic moment and spin; therefore, the scattering length can vary substantially within a given sample, due to the coexistence of several isotopic species and the different magnetic moment orientations within a given isotope. These substantial mean squared fluctuations of b ultimately cause a usually significant incoherent neutron cross-section. Although this represents a severe problem when the experiment aims at investigating the collective dynamics of the sample, it is a valuable resource when the focus is instead on the single-molecule dynamics, as the incoherent scattering directly sheds insight into it. For instance, the almost entirely incoherent neutron cross-section of hydrogen atoms makes neutron scattering measurements on molecular systems containing many of such atoms a unique probe of the single-molecule dynamics.

4.9.2. *Advantages of INS*

Resolution function

A great advantage of *INS* instruments is the narrow and sharp energy resolution function. Most importantly, the energy resolution of *INS* measurements can be tailored to the specific experimental needs, although the refinement of the resolution shape unavoidably shrinks the dynamic range covered. In practice, a resolution function as narrow as a few tens of μeV FWHM or less can be achieved using cold neutrons as a probe. Although new concept *IXS* spectrometers hold the promise of a drastic improvement of both resolution width (sub-meV) and spectral contrast, the performance of quasi-elastic neutron scattering instruments (see, e.g. Bée, 1988) is unlikely to be ever matched by *IXS* spectrometers.

Q-decay of the cross-section

Due to the highly localized interaction of neutrons with the target nuclei, no significant Q-decay of the scattering length occurs up to Q values in the reciprocal order of magnitude of the nuclear size. This value exceeds by nearly three orders of magnitude of the Q value at which the *IXS* cross-section typically halves its $Q = 0$ value. Consequently, the use of neutrons is imperative for extremely high Q measurements probing the so-called single particle or impulse approximation regime (Silver and Sokol, 2013), which, therefore, have been for decades a peculiar task of deep inelastic neutron scattering (DINS) (Sears, 1984). Although the literature documents a few successful attempts to reach the impulse approximation with *IXS* (Monaco, Cunsolo *et al.*, 2002), the maximum Q reached by these investigations is still smaller than the range covered by *DINS* by more than two orders of magnitude.

Absorption

At the incident energies where *IXS* spectrometers are typically operated, the photoelectric absorption is the leading cause of the intensity attenuation. This extinction drastically reduces the Thomson scattering, thus limiting the efficiency of *IXS* measurements on high

Z materials, as the photoelectric absorption coefficient behaves as $\mu \propto Z^4$. Conversely, the absorption cross-section of neutrons is, in most cases, relatively small, which makes neutrons an ideal, nondestructive probe into bulk properties of materials. For this reason, neutron scattering provides a spectroscopic tool well-suited to investigate biological systems or other samples prone to radiation damage (Rheinstadter, Ollinger *et al.*, 2004). Furthermore, the higher penetration depth of neutron beams becomes an advantage in some large volume high-pressure applications, as it allows the use of more compact high-pressure cells made out of a single block of material, with no need of transparent — e.g. diamond or sapphire — windows usually required for X-Ray measurements. This "windowless option" makes the design of large volume high-pressure cells more straightforward, and drastically reduces the risk of leaks.

Single-particle dynamics

As mentioned, the neutron cross-section of hydrogen is substantially higher than that of other atomic species and is almost entirely incoherent. This feature makes neutron scattering a unique tool to probe the single-particle diffusive motions in hydrogen-containing compounds and hydration patterns in proteins and other macro-molecules (Middendorf, 1984).

Contrast variation

When the sample is embedded in the cavities of a porous material or floating in a medium, it is often useful to improve the contrast of the measurement, that is the difference between the scattering from the sample and the background intensity from the confinement medium. Contrast optimisation is more problematic for *IXS* measurements as the X-Ray cross-section depends on the atomic number Z of the target sample, which can only be changed by altering the chemical species of the sample. Conversely, with the neutron scattering cross-section being isotope-specific, the scattering from a composite sample can be manipulated by changing the isotopic species (i.e. the atomic mass) of individual sample components without altering their chemical properties. This prerogative suggests modulating the cross-section using an isotopic substitution to optimize the contrast (Finney and Soper, 1994).

Isotopic substitution

The isotope specificity of the neutron cross-section comes in handy when performing measurements with the 'isotopic substitution' method. These are parallel *INS* measurements on chemically equivalent systems in which changes of the sample isotopic species can be used to single out the partial scattering contributions from individual atomic species in a molecular sample or a mixture (Finney and Soper, 1994).

Magnetic scattering

Neutron scattering is the most effective and versatile experimental tool for studying the microscopic properties of magnetic materials. The magnetic interaction is smaller than the nuclear one; however, it is active over longer distances and, away from the neutron capture resonance, the probability of magnetic scattering events is comparable with the one of inelastic nuclear scattering. The main contribution to magnetic scattering arises from the neutron interaction with the total dipole magnetic moment of the atomic electrons. Magnetic neutron scattering includes several specialized techniques designed for specific studies or particular classes of materials. Among these are magnetic reflectometry investigations of surfaces, interfaces, and multilayers, small-angle scattering of large-scale structures and so forth.

Polarization analysis

In general the scattering of cold and thermal neutrons depends on the relative orientation of the neutron and nuclear spins. If nuclear spins in the sample are randomly oriented, the total scattering contains both an incoherent and a coherent contribution. The first conveys information on *self* density correlations, i.e. correlations involving the positions of the same molecule at different times. The second term arises from correlations between the positions of pairs of molecules at different times, thus being uniquely informative of the collective dynamics. The possibility of selecting the polarization of neutrons offers the opportunity of performing a polarization analysis to ultimately single out these two contributions. Despite some experimental challenges, this method can provide a uniquely informative portray of the dynamic response of a liquid (Squires, 2012; Arbe *et al.* 2020).

References

Alatas, A., Leu, B., Zhao, J., Yavaş, H., Toellner, T. & Alp, E. 2011. *Nucl. Instrum. Meth. Phys. Res. A*, **649**, 166.

Ando, M., Bailey, D. & Hart, M. 1978. *Acta Crystallogr. A*, **34**, 484.

Arbe, A., Nilsen, G. J., Stewart, J. R., Alvarez, F., Garcıa Sakai, V., & Colmenero. J. 2020. *Phys. Rev. Research*, **2**, 022015(R).

Baron, A. Q. R., Tanaka, Y., Miwa, D., Ishikawa, D., Mochizuki, T., Takeshita, K., Goto, S., Matsushita, T., Kimura, H., Yamamoto, F. & Ishikawa, T. 2001. *Nucl. Instrum. Meth. Phys. Res. A*, **467**, 627.

Brockhouse, B. N. & Stewart, A. T. 1955. *Phys. Rev.*, **100**, 756.

Cai, Y. Q., Coburn, D. S., Cunsolo, A., Keister, J. W., Honnicke, M. G., Huang, X. R., Kodituwakku, C. N., Stetsko, Y., Suvorov, A., Hiraoka, N., Tsuei, K. D. & Wille, H. C. 2013. *J. Phys. Conf. Ser.*, **425**, 202001.

Cunsolo, A., Ruocco, G., Sette, F., Masciovecchio, C., Mermet, A., Monaco, G., Sampoli, M. & Verbeni, R. 1999. *Phys. Rev. Lett.*, **82**, 775.

Faigel, G., Siddons, D. P., Hastings, J. B., Haustein, P. E., Grover, J. R., Remeika, J. P. & Cooper, A. S. 1987. *Phys. Rev. Lett.*, **58**, 2699.

Graeff, W. & Materlik, G. 1982. *Nucl. Instrum. Meth. Phys. Res.*, **195**, 97.

Krisch, M. 2003. *Journal of Raman Spectroscopy*, **34**, 628.

Lovesey, S. W. 1984.

Masciovecchio, C., Bergmann, U., Krisch, M., Ruocco, G., Sette, F. & Verbeni, R. 1996a. *Nucl. Instrum. Meth. Phys. Res. B*, **117**, 339.

Masciovecchio, C., Bergmann, U., Krisch, M., Ruocco, G., Sette, F. & Verbeni, R. 1996b. *Nucl. Instrum. Meth. B*, **117**, 339.

Shirane, G., Shapiro, S. M. & Tranquada, J. M. 2002. *Neutron Scattering with a Triple-Axis Spectrometer: Basic Techniques*, Cambridge University Press.

Shvyd'ko, Y., Stoupin, S., Shu, D., Collins, S. P., Mundboth, K., Sutter, J. & Tolkiehn, M. 2014. *Nat. Commun.*, **5**, 4219.

Shvyd'ko, Y. V., Lerche, M., Kuetgens, U., Ruter, H. D., Alatas, A. & Zhao, J. 2006. *Phys. Rev. Lett.*, **97**, 235502.

Shvyd'ko, Y. 2004. *X-Ray Optics — High-Energy-Resolution Applications, Optical Science*, Berlin/Heidelberg/New York, Springer.

Sinn, H., Alp, E. E., Alatas, A., Barraza, J., Bortel, G., Burkel, E., Shu, D., Sturhahn, W., Sutter, J. P., Toellner, T. S. & Zhao, J. 2001. *Nucl. Instrum. Meth. Phys. Res. A*, **467–468**, 1545.

Squires, G. L. 2012. *Introduction to the Theory of Thermal Neutron Scattering*, Cambridge university press.

Suvorov, A., Cai, Y. Q., Sutter, J. P. & Chubar, O. Partially Coherent Wavefront Propagation Simulations for Inelastic X-Ray Scattering Beamline Including Crystal Optics. SPIE Optical Engineering + Applications, 2014. International Society for Optics and Photonics, 92090H.

Toellner, T. S., Alatas, A. & Said, A. H. 2011. *J. Synchrotron Radiat.*, **18**, 605.

Verbeni, R., Sette, F., Krisch, M. H., Bergmann, U., Gorges, B., Halcoussis, C., Martel, K., Masciovecchio, C., Ribois, J. F., Ruocco, G. & Sinn, H. 1996. *J. Synchrotron Radiat.*, **3**, 62.

Willis, B. T. M. & Carlile, C. J. 2017. *Experimental Neutron Scattering*, Oxford University Press.

Windsor, C. G. 1981. *Pulsed Neutron Scattering/C.G. Windsor*, London, Taylor & Francis.

Yabashi, M., Tamasaku, K., Kikuta, S. & Ishikawa, T. 2001. *Rev. Sci. Inst.*, **72**, 4080.

Zachariasen, W. 1967. *Acta Crystallogr.*, **23**, 558.

Chapter 5

From the Mori–Zwanzig formalism
to the lineshape model

After discussing theoretical and experimental aspects of the *IXS* method, we now focus on the choice of an appropriate model for the measured *IXS* signal and, primarily, for the variable it directly relates to, namely $S(Q,\omega)$. The exact analytical form of $S(Q,\omega)$ is known in two extreme dynamic regimes only: at the lowest Q,ω's (hydrodynamic regime) or at the highest ones (single-particle regime). In between these two limits, density fluctuations become strongly coupled with internal degrees of freedom of the fluid; the effects of this coupling on $S(Q,\omega)$ are often subtle and currently lack an analytical prediction. The absence of a firm lineshape theory appears especially penalizing as the spectral evolution along this crossover is uniquely informative of the various dynamical events occurring in a system at mesoscopic scales. Currently, the spectrum is mostly described by approximate models, which are often phenomenological and affected by inherent limitations such as their pertinence to the specific dynamic windows or thermodynamic conditions of the system. Also, the comparison between different analytical models is not always straightforward, and an effort toward a unified formalism is highly desirable (Bafile *et al.*, 2006).

Before describing the most popular lineshape models in detail (Chapter 6), it is useful to introduce a general formalism which many of these models can be derived from, customarily referred to as the memory function formalism (Zwanzig, 1965, Zwanzig, 1961, Mori, 1965b). This is the main objective of this chapter.

5.1. Some general considerations on the memory function formalism (Berne and Pecora, 1976)

In Chapter 2, we have discussed some relevant properties of the time correlation functions of a single dynamic variable $A(t)$, also emphasising their relevance to spectroscopic measurements. In many cases, the behavior of the system at hand is conveniently represented using a set of time-dependent properties rather than just one. Although the actual choice of such a set might not be obvious, let us assume that has been identified in a k-tuple of variables $A_i(t)$ with $i = 1, \ldots, k$. These properties can be seen as the components of the column vector:

$$\boldsymbol{A}(t) = \begin{bmatrix} A_1(t) \\ \vdots \\ A_k(t) \end{bmatrix}, \tag{5.1}$$

whose time-dependence is ruled by the Liouville vector equation:

$$\frac{d\boldsymbol{A}(t)}{dt} = i\mathfrak{L}\boldsymbol{A}. \tag{5.2}$$

As an obvious generalisation of what discussed for the single variable case, the solution of the above equation assumes the form $\boldsymbol{A}(t) = \exp(i\mathfrak{L}t)\boldsymbol{A}$. Again, one can use the formal expression of the exponential of an operator and write $\boldsymbol{A}(t) = [\sum_{n=1}^{\infty}(i\mathfrak{L}t)^n/n!]\boldsymbol{A}$. Unless further approximations are made — assuming, for instance, that the linear ($n = 1$) term of the series is dominating — reducing the solution of Eq. (5.2) to a simple analytical form is prohibitive.

In this chapter we will rather focus on the correlation function associated to $\boldsymbol{A}(t)$, for which we use here the shorthand notation $\boldsymbol{C}(t) = \boldsymbol{C}_{AA}(t)$. We will see that a considerable simplification in the search for an explicit expression of $\boldsymbol{C}(t)$ comes from the possibility of using a linear approximation in the equation defining its time dependence, namely:

$$\frac{d\boldsymbol{C}(t)}{dt} = \mathcal{K}\boldsymbol{C} \tag{5.3}$$

where $\mathcal{K}$ is a matrix composed of time-independent terms.

We will show further below that the linear form of Eq. (5.3) is retrieved when a limited number of slowly fluctuating variables can be used to describe the properties of the system. If these variables cannot be identified, a robust method to analytically handle the the explicit expression of the correlation function can be found in the framework of the Mori–Zwanzig formalism through a continued fraction expansion of the memory kernel. This formalism relies on the use of projection operators, which presents intriguing analogies with the Dirac formalism of the "bras" and "kets" and the related notion of projectors in quantum mechanics (Sakurai and Commins, 1995).

This chapter will briefly illustrate how this formalism ultimately enables to build up a model for the spectral shape through a suitable *ansatz* for the memory function.

As further elucidated in Chapter 6, despite the undoubted success of this formalism in consistently accounting for an increasing number of experimental results, the memory function-based modeling of the lineshape is phenomenological in spirit, and its accuracy mostly rests on appropriate yet somehow arbitrary *ansatz* on the time-dependent memory function. Nonetheless, the physical soundness of memory function-based lineshape models can be corroborated by the superimposition of the sum rules or other constraints to the values of relevant transport and thermodynamic parameters of the fluid.

5.2. The Generalized Langevin Equation (Berne and Pecora, 1976)

We introduce here some relevant aspects of the Mori–Zwanzig formalism, and discuss how correlation functions and related spectra can be consistently described in this framework.

(1) Defining a scalar product

As mentioned above, we here assume that the description of the physical problem relevant to a many-body system requires a set of stochastic variables, rather than only one, and this can be represented by the column vector in Eq. (5.1).

This vectorial representation suggests some correspondence with the ordinary vector space, which becomes more precise once we introduce the notion of scalar product, hereafter denoted with the symbol "$(\ldots,\ldots)$".

Such a product must be a complex number fulfilling the following general properties:

(1) $(\boldsymbol{A}, \boldsymbol{A}^{\dagger}) \in \mathbb{R}$ and $0 \leq (\boldsymbol{A}, \boldsymbol{A}^{\dagger}) \leq \infty$; if $(\boldsymbol{A}, \boldsymbol{A}^{\dagger}) = 0 \Rightarrow \boldsymbol{A} \equiv 0$
(2) $(\boldsymbol{A}, \boldsymbol{B}^{\dagger})^{*} = (\boldsymbol{B}, \boldsymbol{A}^{\dagger})$
(3) $(\sum_{k} c_{k}\boldsymbol{A}_{k}, \boldsymbol{B}^{\dagger}) = \sum_{k} c_{k}(\boldsymbol{A}_{k}, \boldsymbol{B}^{\dagger})$
(4) $(\boldsymbol{A}, (\sum_{k} c_{k}\boldsymbol{B}_{k})^{\dagger}) = \sum_{k} c_{k}^{*}(\boldsymbol{A}, \boldsymbol{B}_{k}^{\dagger})$,

where the Hermitian conjugate of the column vector $\boldsymbol{A}$ introduced above is the row vector $\boldsymbol{A}^{\dagger} = [A_{1}^{*}, \ldots, A_{k}^{*}]$, where A^{*} is the complex conjugate of the variable A.

If $(\boldsymbol{A}, \boldsymbol{B}^{\dagger}) = 0$, the variables $\boldsymbol{A}$ and $\boldsymbol{B}$ are said to be orthogonal, while if $(\boldsymbol{A}, \boldsymbol{A}^{\dagger}) = \boldsymbol{I}$, with $\boldsymbol{I}$ being the identity matrix, then $\boldsymbol{A}$ forms an orthonormal set of variables.

The properties required for a scalar product, as defined above, are fulfilled by the correlation function, which in this k-dimensional space is defined as:

$$C_{AB} = \langle \boldsymbol{AB}^{\dagger} \rangle \equiv (\boldsymbol{A}, \boldsymbol{B}^{\dagger}). \tag{5.4}$$

The correlation matrix $\boldsymbol{C}_{AB}$ is a $k \times k$ matrix whose generic jmth element is the correlation between the jth component of the column vector $\boldsymbol{A}$ and the mth component of the row vector $\boldsymbol{B}^{\dagger}$, i.e. $C_{AB}^{jm} \equiv \langle A_{j}B_{m}^{*} \rangle$. To frame the concept in a broader context, the identification of the correlation with the scalar product is supported by the application of the linear response theory to classical systems (Mori, 1965a).

As mentioned in the previous section, the very notion of "orthogonality" now acquires a statistical nuance: two or more sets of variables are dubbed orthogonal when statistically uncorrelated (see, e.g., Berne, 1977).

Let us now assume that the k-tuple $A_1, \ldots, A_k$ is composed of mutually independent variables, which means that a single variable A_i cannot be expressed as a linear combination of the remaining $k - 1$. In close analogy with ordinary space vectors, starting from an ensemble of k independent variables, it is always possible to build up an orthonormal set, i.e. a set for which $\langle A_j A_m^* \rangle = \delta_{jm}$ with δ_{jm} being the Kronecker delta.

It is implicit in the above arguments that the notion of statistical independence does not imply orthogonality, i.e. statistically independent variables are not necessarily orthogonal, *i.e.* they do not necessarily have vanishing correlation.

Given the notion of scalar product defined above, it seems natural to infer that if some dynamic variables are statistically independent, unless orthogonal, they have a non-vanishing mutual projection. Again, in analogy with the standard vector space, one can define the projection operator in the subspace of $\boldsymbol{A}$ as:

$$\mathfrak{p} = (\ldots, \boldsymbol{A}^\dagger) \cdot (\boldsymbol{A}, \boldsymbol{A}^\dagger)^{-1} \cdot \boldsymbol{A}. \tag{5.5}$$

It readily appears that this operator acts on a generic stochastic variable $\boldsymbol{B}$ by projecting it in the subspace parallel to $\boldsymbol{A}$. Here $(\boldsymbol{A}, \boldsymbol{A}^\dagger)^{-1}$ is the inverse of the matrix $(\boldsymbol{A}, \boldsymbol{A}^\dagger)$, while the dots represent the ordinary product between matrices. Given this definition, the vector $\boldsymbol{B}$ projected on $\boldsymbol{A}$ reads $\mathfrak{p}\boldsymbol{B} = (\boldsymbol{B}, \boldsymbol{A}^\dagger) \cdot (\boldsymbol{A}, \boldsymbol{A}^\dagger)^{-1} \cdot \boldsymbol{A}$, where $(\boldsymbol{B}, \boldsymbol{A}^\dagger) \cdot (\boldsymbol{A}, \boldsymbol{A}^\dagger)^{-1}$ is the $\boldsymbol{B}$ component along $\boldsymbol{A}$.

If $\mathfrak{p}\boldsymbol{B} = 0$, $\boldsymbol{B}$ has a vanishing component along $\boldsymbol{A}$, i.e. it is orthogonal to it. We can thus consider the subspace of stochastic variables orthogonal to (uncorrelated with) $\boldsymbol{A}$, and introduce the operator of projection in such an orthogonal subspace:

$$\mathfrak{q} = \boldsymbol{I} - \mathfrak{p}. \tag{5.6}$$

Given the variable $\boldsymbol{A}$, we can consider its time-propagated counterpart $\boldsymbol{A}(t) = \exp(i\mathfrak{L}t)\boldsymbol{A}$ and its time correlation matrix:

$$\boldsymbol{C}_{AA}(t) = (\boldsymbol{A}(t), \boldsymbol{A}^\dagger) \tag{5.7}$$

The value of this matrix at the time $t = 0$ is often denoted as $\boldsymbol{C}_{AA}(0) = (\boldsymbol{A}, \boldsymbol{A}^\dagger) = \beta^{-1}\chi$, where the matrix χ referred to as the static susceptibility matrix. Notice that χ is Hermitian as $(A, A^\dagger)^\dagger = (A, A^\dagger)$. Upon approaching some thermodynamic points of instability, as for instance a gas liquid-critical point, the susceptibilities relating to the fluctuation of some order parameters tend to diverge.

(2) The generalized Langevin equation

Although a rigorous demonstration is not given here (see (Berne and Pecora, 1976) for details), the use of the projection operators in Eqs. (5.5) and (5.6) combined with suitable analytic manipulations eventually transforms Eq. (5.2) into the following equation:

$$\frac{dA(t)}{dt} = i\Omega \cdot A(t) - \int_0^t d\tau K(\tau) \cdot A(t - \tau) + f(t). \qquad (5.8)$$

The one above is customarily referred to as the generalized Langevin equation, as it generalizes the ordinary Langevin one, which lacks the integral term.

The generalized Langevin equation contains a few new variables:

- **The fluctuating random force $f(t)$**
  This term can be expressed using both the Liouville and the projection operators as:

$$f(t) = \exp(iq\mathcal{L}t)q(i\mathcal{L}A) = \exp(iq\mathcal{L}t)q\dot{A} \qquad (5.9)$$

where we made use of the Liouville Equation $\dot{A}(t) = i\mathcal{L}A$. From the properties of the projection operator q, it follows that $(f(t), A^\dagger) = 0$, *i.e.* the force is orthogonal to (uncorrelated with) $A = A(0)$. Notice that the time evolution of the random force is determined by the propagator $\exp(iq\mathcal{L}t)$.[1]

---

[1]In theoretical physics the expression of the time-propagator as an exponential of either the Liouvillian (classical case) or the Hamiltonian (quantum case) is rather common. However, the inclusion in the exponent of the projection operator is a prerogative of the Mori–Zwanzig formalism, which unfortunately makes the analytical derivation of the random force more cumbersome.

- **The memory matrix $K(t)$**
  The memory of the system is embodied in the kernel:

$$K(t) = (f(t), f^\dagger) \cdot (A, A^\dagger)^{-1}, \qquad (5.10)$$

which is hereafter referred to as the memory matrix, or, in the single variable case $(k = 1)$, the memory function. The memory integral in Eq. (5.8) is compliant with the causality principle, as the integration runs over times preceding t, *i.e.* it covers the $\tau \in (0, t)$ interval, thus describing the ability of the system to keep the memory of the past.

- **The frequency matrix**
  The proper frequency of the system is described by the frequency matrix:

$$\Omega = (\mathcal{L}A, A^\dagger) \cdot (A, A^\dagger)^{-1} = (-i\dot{A}(t), A^\dagger) \cdot (A, A^\dagger)^{-1}, \quad (5.11)$$

which characterizes the oscillatory behavior of $A(t)$. This term embodies the non-dissipative, or elastic-like portion of the dynamic response, while dissipative effects are accounted for by the kernel $K(\tau)$.

As mentioned, Eq. (5.8) generalizes the ordinary Langevin Equation which describes, for instance, the Brownian motion of a particle. In this simple case, the description of the time-dependent velocity hinges on two main assumptions: (i) the random force has a Gaussian distribution, and (ii) its correlation function decays very rapidly to the extent of being consistently approximated by a δ-function of time. The Gaussian assumption is quite reasonable, as the Brownian motion is accomplished a particle much more massive than surrounding molecules it collides with. Consequently, the motion results from many successive collisions, which is a prerequisite for the validity of the central limit result. This also justifies the second assumption, since memory effects persist only for timescale comparable to those featuring erratic molecular movements, *i.e.* much more rapid than Brownian motions. In principle, the generalized Langevin equation does not rest on any of these assumptions (Kubo, 1966).

Notice that the scalar product of both members of Eq. (5.8) with $\mathbf{A}^\dagger$ yields:

$$\frac{d\mathbf{C}(t)}{dt} = i\mathbf{\Omega} \cdot \mathbf{C}(t) - \int_0^t \mathbf{K}(\tau) \cdot \mathbf{C}(t-\tau)d\tau, \qquad (5.12)$$

where we used the shorthand notation $\mathbf{C}(t) = \mathbf{C}_{AA}(t)$. Indeed, after performing the scalar product, the random force disappears due to its orthogonality to $\mathbf{A}$, although it still influences — through its normalized correlation $\mathbf{K}(t)$ — the time-dependence of $\mathbf{C}(t)$.

The one introduced above is customarily referred to as the memory equation. This equation rules the time evolution of an equilibrium average as the correlation function. In this respect, it is not surprising that rapid random variations typical of Brownian phenomena are not represented in this equation, which describes a smooth regular motion. Conversely, Eq. (5.8) also contains an erratically changing rapid variable, *i.e.* the random force, which in principle accounts for "non-linear effects, the initial transient process, and fluctuations" (Mori, 1965a) affecting the time evolution of $\mathbf{A}(t)$.

5.3. Identifying a set of slow variables (Keyes, 1977)

The solution of the generalized Langevin equation in Eq. (5.8) becomes more straightforward when the physical problem at hand can be described by a limited number of "slow" variables, i.e. variables which fluctuate much more slowly than all the other state variables of the system.

Although there is no rigorous criterion for the selection of the right set of slow variables, we shall assume that, thanks to the investigator judgment and intuition, such statistically independent variables $A_i(t)$ (with $i = 1, \ldots, k$) are identified. Let us further assume that these variables form a complete set, meaning, any other conceivable slow variables of the system can be written as a linear combination of them.

Since all variables $A_i(t)$ decay over timescales much slower than any other dynamic variables of the system, one can assume

$A_i(t) \approx A_i(0)$; this implies that the random forces, which are orthogonal to $A_i(0)$, are also orthogonal to $A_i(t)$ for all considered t. Their orthogonality to the complete set, $A_i(t)$, prevents the random forces from having "slow" components, and leaves them confined inside the subspace of "fast" or rapidly fluctuating variables.

Aside from the rather vague character of these arguments, it can be shown immediately that the identification of such a complete set of slow variables substantially simplifies the solution of the generalized Langevin equation. In fact, if the forces are "fast", their normalized autocorrelation functions, i.e. the memory kernel (see Eq. (5.10)), supposedly relaxes within a timescale much shorter than the one characterizing the evolution of slow variables. Pushing this argument to the extreme, one can assume that the memory becomes reasonably well approximated by a δ-function of time (Markov approximation). Hence:

$$K(\tau) = \Gamma_0 \delta(t), \tag{5.13}$$

where Γ_0 is often referred to as the relaxation matrix. Notice that the above approximation for the memory function is appropriate for a system which loses memory instantaneously, or, equivalently, has no memory at all.

Under this condition, the convolution integral in Eq. (5.8) transforms into a product:

$$\int_0^t d\tau \, K(\tau) \cdot A(t - \tau) = \Gamma_0 \cdot \int_0^\infty d\tau \, A(t - \tau) \delta(\tau) = \Gamma_0 \cdot A(t).$$
$$\tag{5.14}$$

and, correspondingly, the generalized Langevin equation reduces to the following set of interconnected linear differential equations:

$$\frac{dA(t)}{dt} = (i\Omega - \Gamma_0) \cdot A(t) + f(t). \tag{5.15}$$

The equation for the correlation matrix can be simply obtained by multiplying both members of the above equation by $A^\dagger$ and performing the equilibrium average. Explicitly:

$$\frac{d\boldsymbol{C}(t)}{dt} = (i\boldsymbol{\Omega} - \boldsymbol{\Gamma}_0) \cdot \boldsymbol{C}(t), \qquad (5.16)$$

which defines a set of independent linear differential equations, whose solution can be easily tackled analytically. Upon integrating Eq. (5.13), one obtains

$$\boldsymbol{C}(t) = \exp[(i\boldsymbol{\Omega} - \boldsymbol{\Gamma}_0)t] \qquad (5.17)$$

which shows that $\boldsymbol{C}(t)$ experiences slow damped oscillations at frequency $\boldsymbol{\Omega}$, where the damping is determined by $\boldsymbol{\Gamma}_0$.

The low Q behavior of frequency and damping

The identification of the slow variables becomes straightforward whenever the Hamiltonian has some macroscopic constants of motion, say the set of extensive variables $\boldsymbol{A}^{\mathrm{EXT}}(t) = \boldsymbol{A}^{\mathrm{EXT}}$.

We recall here that an extensive property can be written as the sum of the properties of the N-atoms in the (monatomic) system as $\boldsymbol{A}^{\mathrm{EXT}} = \sum_{j=1}^{N} \boldsymbol{a}_j(t)$. For instance, $\boldsymbol{A}^{\mathrm{EXT}}$ can be the energy, the mass or the momentum of the system.

One can define the microscopic density of such a set i.e. $\boldsymbol{A}(\boldsymbol{r}, t) = \sum_{j=1}^{N} \boldsymbol{a}_j(t) \delta[\boldsymbol{r} - \boldsymbol{R}_j(t)]$.[2] It is instructive to consider the time derivative of the Fourier transform, $\boldsymbol{A}(Q, t) = \sum_{j=1}^{N} \boldsymbol{a}_i(t) \exp[i\boldsymbol{Q} \cdot \boldsymbol{R}_j(t)]$, which reads as:

$$\dot{\boldsymbol{A}}(\boldsymbol{Q}, t) = iQ \sum_{j=1}^{N} [\widehat{\boldsymbol{Q}} \cdot \boldsymbol{v}_j(t)] \boldsymbol{a}_i(t) \exp[i\boldsymbol{Q} \cdot \boldsymbol{R}_j(t)]$$

$$+ \sum_{j=1}^{N} \dot{\boldsymbol{a}}_i(t) \exp[i\boldsymbol{Q} \cdot \boldsymbol{R}_j(t)].$$

---

[2]The interpretation of $\boldsymbol{A}(\boldsymbol{r}, t)$ as the "density of" the extensive variable $\boldsymbol{A}^{\mathrm{EXT}}$ is perhaps more evident if one considers that $\int_V d\boldsymbol{r}\, \boldsymbol{A}(\boldsymbol{r}, t) = \sum_{j=1}^{N} \dot{\boldsymbol{a}}_i(t) = \boldsymbol{A}^{\mathrm{EXT}}$; by taking the derivative of first and last members of this identity chain with respect to the volume, one has $\boldsymbol{A}(\boldsymbol{r}, t) = d\boldsymbol{A}^{\mathrm{EXT}}/dV$.

Here, $\boldsymbol{v}_i(t)$ is the velocity of the jth atom. For low Q's, one can use the approximation $\exp[i\boldsymbol{Q} \cdot \boldsymbol{R}_j(t)] \approx 1 + iQ\widehat{\boldsymbol{Q}} \cdot \boldsymbol{R}_j(t)$, which yields:

$$\dot{\boldsymbol{A}}(\boldsymbol{Q}, t) \approx iQ \left\{ \sum_{j=1}^{N} \{ [\widehat{\boldsymbol{Q}} \cdot \boldsymbol{v}_j(t)] \boldsymbol{a}_i(t) + \dot{\boldsymbol{a}}_i(t) \} \widehat{\boldsymbol{Q}} \cdot \boldsymbol{R}_j(t) \right\} \qquad (5.18)$$

where we neglected the $O(Q^2)$ terms and considered that, since $\boldsymbol{A}^{\text{EXT}} = \sum_{j=1}^{N} \boldsymbol{a}_i(t)$ is a constant of motion, then $\sum_{j=1}^{N} \dot{\boldsymbol{a}}_j(t) = 0$.

From the equation above, it appears that the time derivative of $\boldsymbol{A}(\boldsymbol{Q}, t)$ vanishes linearly with Q in the low limit. Notice that out of the $Q = 0$ limit the intensive variable $\boldsymbol{A}(\boldsymbol{Q}, t)$ is not conserved.

Let us now consider the implication of Eq. (5.18) for the time evolution of the Q-generalized correlation function $\boldsymbol{C}_{\boldsymbol{Q}}(t) = (\boldsymbol{A}(\boldsymbol{Q}, t), \boldsymbol{A}^\dagger(\boldsymbol{Q}))$, as expressed in Eq. (5.14). From a closer look at the various terms of such an equation one can deduce that:

(1) The first member of Eq. 5.14 is

$$\dot{\boldsymbol{C}}_{\boldsymbol{Q}}(t) = (\dot{\boldsymbol{A}}(\boldsymbol{Q}, t), \boldsymbol{A}^\dagger(\boldsymbol{Q})) \qquad (5.19)$$

and contains $\dot{\boldsymbol{A}}(\boldsymbol{Q}, t)$ once; therefore, at low Q values, this term vanishes as $\dot{\boldsymbol{A}}(\boldsymbol{Q}, t)$ does, *i.e.* as an $O(Q)$ infinitesimal.

(2) The Q-generalized frequency matrix

$$\boldsymbol{\Omega}_{\boldsymbol{Q}} = -i(\dot{\boldsymbol{A}}(\boldsymbol{Q}, t), \boldsymbol{A}^\dagger(\boldsymbol{Q})) \cdot (\boldsymbol{A}(\boldsymbol{Q}), \boldsymbol{A}^\dagger(\boldsymbol{Q}))^{-1},$$

also contains $\dot{\boldsymbol{A}}(\boldsymbol{Q}, t)$ once, thus being an $O(Q)$ infinitesimal for low Q values.

(3) Finally, the memory matrix

$$\boldsymbol{K}_{\boldsymbol{Q}}(t) = (\boldsymbol{f}(\boldsymbol{Q}, t), \boldsymbol{f}^\dagger(\boldsymbol{Q})) \cdot (\boldsymbol{A}(\boldsymbol{Q}), \boldsymbol{A}^\dagger(\boldsymbol{Q}))^{-1} \qquad (5.20)$$

with $\boldsymbol{f}(\boldsymbol{Q}, t) = \exp(i\mathfrak{q}\mathfrak{L}t)\mathfrak{q}\dot{\boldsymbol{A}}(\boldsymbol{Q}, t)$; hence, $\boldsymbol{K}_{\boldsymbol{Q}}(t)$ contains the product $\dot{\boldsymbol{A}}(\boldsymbol{Q}, t)\dot{\boldsymbol{A}}^\dagger(\boldsymbol{Q}, 0)$, thus being, at low Q's, an $O(Q^2)$ infinitesimal, and so is the relaxation matrix appearing in Eq. (5.14).

From the arguments above, it follows that at low Q and within the validity of the Markovian approximation, the right-hand side of Eq. (5.15) reduces to the sum of: a term $O(Q)$ describing slow oscillatory movements of $C(Q,t)$ and a $O(Q^2)$ damping term determined by the relaxation matrix.

Hence, at low Q's, the damping matrix, being an $O(Q^2)$ infinitesimal, vanishes faster than the frequency, which is only $O(Q)$.

In summary, within a Markovian approximation for the memory, at low Q the slow oscillations experienced by the correlation become long living. The two acoustic modes in the so called Rayleigh Brillouin (see Chapter 6) spectrum provide an example of these long living oscillation modes.

5.4. Beyond the Markov approximation (Balucani and Zoppi, 1994)

Summarizing, we showed in the last section that the identification of a complete set of slow variables considerably simplifies the solution of the Langevin equation. Unfortunately, "nature is not always so gentle (or our understanding of the problem so deep)" (Balucani and Zoppi, 1994), and often a set of slow variables is not easily identifiable. One can cope with this difficulty by desisting from the search of a complete set of slow variables and tackling analytically non-Markovian effects on the memory function.

The non-Markovian short-time behaviour of the memory function can in principle be unraveled by performing a series expansion.

Without explicit derivation (the interested reader is referred to (Balucani and Zoppi, 1994)), we emphasize here that the coefficients of such an expansion, i.e. the time-derivatives of the memory matrix computed in $t = 0$, are linked to the nth order frequency moments of the spectrum $C(\omega)$:

$$\langle \omega^n \rangle = \frac{1}{i^n} \left. \frac{d^n C(t)}{dt^n} \right|_{t=0} = \int_{-\infty}^{+\infty} d\omega \, \omega^n C(\omega). \tag{5.21}$$

The fact that the short-time behavior of the memory function is directly related to spectral moments is particularly fortunate because spectral moments are in principle derivable as equilibrium averages. Therefore, the various $t = 0$ derivatives of the memory function governing such an expansion are, in principle amenable to an explicit form, although the calculation of moments may be demanding and their physical interpretation not straightforward.

An alternative pathway to handle non-Markovian effects is given by the so-called continued fraction expansion of the memory function, and it is here illustrated succinctly.

The memory function equation in Eq. (5.12) is a vectorial equation which defines a set of integrodifferential equations of the Volterra kind (Vladimirov, 1984). As a first step, the solution of this system of equations can be pursued by a Laplace transform, which converts the convolution integral therein into a product. After this manipulation, Eq. (5.12) transforms to:

$$\widetilde{C}(z) = [zI - i\Omega + \widetilde{K}(z)]^{-1} \cdot C(0), \qquad (5.22)$$

where, according to the usual notation $\widetilde{C}(z)$ and $\widetilde{K}(z)$ are the Laplace transforms of the correlation and memory matrices, respectively.

At this stage, it can be recognized that the random force in the Langevin equation is, in turn, a stochastic variable which obey to an equation of motion of the Liouville type. Specifically, it obeys a Liouville-type equation in which the Liouville operator is replaced by its counterpart operating in the orthogonal space $\mathfrak{q} = (I - \mathfrak{p}) \cdot \mathcal{L}$, as evident from Eq. (5.8). Thus, it is useful to introduce the new projection operators:

$$\mathfrak{p}_1 = (\dots, f^\dagger) \cdot (f, f^\dagger) \cdot f \quad \mathfrak{q}_1 = I - \mathfrak{p}_1, \qquad (5.23)$$

which can be used to derive the following generalized Langevin equation for the random force:

$$\frac{df(t)}{dt} = i\Omega_1 \cdot f(t) - \int_0^t K_1(\tau) \cdot f(t - \tau)d\tau + f_1(t). \qquad (5.24)$$

Here the new random force acting on $f(t)$ is $f_1(t) = \exp(i\mathfrak{q}_1\mathcal{L}_1 t)i\mathfrak{q}_1\mathcal{L}_1 f$, where $\mathcal{L}_1 = \mathfrak{q}\mathcal{L}$, the new frequency matrix is $\Omega_1 = -i(\dot{f}, f^\dagger) \cdot (f, f^\dagger)^{-1}$ and the memory matrix $K_1(t) = (f_1(t), f_1^\dagger) \cdot (f, f^\dagger)^{-1}$.

Since the generalized Langevin equation determines the time evolution of $f(t)$, an equation similar to Eq. (5.11) describes the time evolution of its normalized correlation, $\boldsymbol{K}(t) = (\boldsymbol{f}(t), \boldsymbol{f}^\dagger) \cdot (\boldsymbol{A}, \boldsymbol{A}^\dagger)^{-1}$, whose Laplace transform is similar to the one previously written for the correlation matrix (Eq. (5.22)), namely:

$$\widetilde{\boldsymbol{K}}(z) = [z\boldsymbol{I} - i\boldsymbol{\Omega}_1 + \widetilde{\boldsymbol{K}}_1(z)]^{-1} \cdot \boldsymbol{K}(0). \tag{5.25}$$

The insertion in the above equation the Laplace transform of $\boldsymbol{K}_1(t)$ (defined before) and the successive anti-transform of the equation, provides a formally exact solution for $\boldsymbol{K}(t)$. Also a formally exact expression can be achieved for the correlation matrix. Indeed, the link between $\widetilde{\boldsymbol{C}}(z)$ and $\widetilde{\boldsymbol{K}}_1(z)$ follows from inserting the above equation into Eq. (5.22), which leads to:

$$\frac{\widetilde{\boldsymbol{C}}(z)}{\boldsymbol{C}(0)} = \left[z\boldsymbol{I} - i\boldsymbol{\Omega} + \left[z\boldsymbol{I} - i\boldsymbol{\Omega}(1) + \widetilde{\boldsymbol{K}}_1(z)\right]^{-1} \cdot \boldsymbol{K}(0)\right]^{-1}. \tag{5.26}$$

The arguments illustrated above suggest following an iterative approach, which, in the single variable case, corresponding to $\Omega = 0$,[3] leads to the following continued fraction expansion:

$$\frac{\tilde{C}(z)}{C(0)} = \left[z + \frac{\Delta_1}{z + \Delta_2/(z + \ldots)}\right] \tag{5.27}$$

which can be used to derive an explicit form for the Laplace transform of the correlation function. Here the values of the parameter $\Delta_n = K_{n-1}(0)$ directly relate to the normalized moments of the $C(\omega)$ spectrum. For the first few orders one has:

$$\Delta_1 = K(0) = \frac{\langle \omega^2 \rangle}{\langle \omega^0 \rangle} \tag{5.28}$$

$$\Delta_2 = K_1(0) = -\frac{\ddot{K}(0)}{K(0)} = \frac{\langle \omega^4 \rangle}{\langle \omega^2 \rangle} - \frac{\langle \omega^2 \rangle}{\langle \omega^0 \rangle} \tag{5.29}$$

---

[3]It can be easily demonstrated (see Berne and Pecora 1976) that the matrix Ω only involves correlations between distinct variables with opposite time reversal properties, thus vanishing in the single variable case.

$$\Delta_3 = K_2(0) = -\frac{\ddot{K}_1(0)}{K_1(0)}$$

$$= \frac{1}{\Delta_2}\left[\frac{\langle\omega^6\rangle}{\langle\omega^2\rangle} - \left(\frac{\langle\omega^4\rangle}{\langle\omega^2\rangle}\right)^2\right]. \tag{5.30}$$

In the remainder of this book, we will consider a second-order truncation of the continued fractions expansion discussed above, where the microscopic density fluctuation is the dynamic variable of interest. The second order memory function for the microscopic density coincides with the first order memory function of the microscopic current i.e. the spectrum of atomic velocities, defined in Chapter 2. This stems from the circumstance that microscopic velocities can be essentially regarded as the time derivatives of microscopic densities, and that an n-order memory function of a generic dynamic variable corresponds to the $n-1$ order memory of its time-derivative.

5.5. From the memory function to the spectral lineshape

In summary, when the only relevant variable is the density fluctuation, its spectrum can be derived using a 2nd order truncation of the continued fraction expansion, which yields:

$$\frac{S(Q,\omega)}{S(Q)} = \frac{1}{\pi}Re\left[\frac{\tilde{F}(Q,z=i\omega)}{S(Q)}\right]$$

$$= \frac{1}{\pi}Re\left[z + \frac{\omega_0^2(Q)}{z + \tilde{m}_L(Q,z)}\right]^{-1}_{z=i\omega}, \tag{5.31}$$

where $F(Q,t)$ is the intermediate scattering function introduced in Chapter 2 (Eq. (2.82)), while $\tilde{m}_L(Q,z)$ is the Laplace transform of the Q-dependent second order memory function $\tilde{m}_L(Q,t)$. In the above equation, we have considered the link between Fourier and Laplace transform (see Eq. (2.68)).

In general, when modeling the lineshape, $S(Q,\omega)$, the physical soundness of the used model can be improved by superimposing to it the correct values of the spectral moments. The formula providing an exact expression for the nth spectral moment $\langle\omega^n\rangle$ is customarily referred to as the nth sum rule. In particular, the fulfilment of the first two (even) sum rules for a classical monatomic fluid leads to $\omega_0^2(Q) = \langle\omega^2\rangle/\langle\omega^0\rangle = Q^2 k_B T/[MS(Q)]$ (see, for instance, Balucani and Zoppi, 1994). The superimposition of higher order sum rules to the spectrum will be discussed in the next chapter, with specific reference to the viscoealstic model.

Eq. (5.31) can be further developed to eventually obtain:

$$\frac{S(Q,\omega)}{S(Q)} = \frac{1}{\pi} \frac{\omega_0^2 m_L'(Q,\omega)}{[\omega^2 - \omega_0^2 - \omega m_L''(Q,\omega)]^2 + \omega^2[m_L'(Q,\omega)]^2}. \quad (5.32)$$

Here $m_L'(Q,\omega)$ and $-m_L''(Q,\omega)$ are, respectively, real and imaginary parts of the Laplace transform of $m_L(Q,t)$, *i.e.*:

$$\tilde{m}_L(Q, z = i\omega) = \int_0^\infty dt\cos(\omega t)m_L(Q,t) - i\int_0^\infty dt\sin(\omega t)m_L(Q,t)$$

$$= m_L'(Q,\omega) - im_L''(Q,\omega).$$

When attempting to describe the spectral profile measured in an *IXS* experiment the use of Eq. (5.29) shifts the focus of the model from $S(Q,\omega)$ to $\tilde{m}_L(Q, z = i\omega)$, or, in the time domain, to $m_L(Q,t)$.

References

Bafile, U., Guarini, E. & Barocchi, F. 2006. *Physical review. E, Statistical, nonlinear, and soft matter physics*, **73**, 061203.

Balucani, U. & Zoppi, M. 1994. *Dynamics of the Liquid State*, Oxford, Clarendon Press.

Berne, B. J. 1977. *In: Statistical Mechanics, Part B: Time-Dependent Processes*. Editor Berne, B. J. Plenum Press, New York, Pag. 233.

Berne, B. J. & Pecora, R. 1976. *Dynamic Light Scattering*, Wiley, New York.

Keyes, T. 1977. *In: Statistical Mechanics, Part B: Time-Dependent Processes*. Editor Berne, B. J. Plenum Press, New York, Pag. 259

Mori, H. 1965. *Prog. Theor. Phys.*, **33**, 423.

Onsager, L. 1931a. *Phys. Rev.*, **37**, 405.

Onsager, L. 1931b. *Phys. Rev.*, **38**, 2265.

Sakurai, J. J. & Commins, E. D. 1995. *Modern Quantum Mechanics*, Revised Edition. AAPT.

Vladimirov, V. S. 1984. *Equations of Mathematical Physics*, Mir, Moscow.

Zwanzig, R. 1965. *J. Chem. Phys.*, **43**, 714.

Chapter 6

A model for the lineshape

After discussing some general aspects of the memory function formalism, and ultimately setting a formal link between the dynamic structure factor $S(Q, \omega)$ and the Laplace transform of the memory function (Eq. 5.28), we now focus on the actual modelling of the latter variable. All models to be discussed in this Chapter strictly refer to a classical description of the spectrum of density fluctuations in which all relevant operators and variables are treated as commuting quantities. Sizable quantum deviations are assumed to only influence the statistical population of states with energy $\hbar\omega$, via the detailed balance principle; specifically, they cause an asymmetry of the lineshape which can be consistently accounted for by the simple insertion of a frequency-dependent prefactor in the model, as discussed at the end of this chapter.

As previously mentioned, the analytical form of $S(Q, \omega)$ is known in two extreme limiting regimes only: the hydrodynamic and the single-particle ones. In the former regime, the response of the system is averaged over such long distances and timescales that the sample can be represented as a continuum, whose dynamic response is probed as an average over many microscopic events.

At the infinitesimal Q, ω values typical of this regime, the Navier–Stokes theory of hydrodynamics can be consistently used to describe the spectrum of density fluctuations of a fluid. This treatment leads to predict the well-known Rayleigh–Brillouin triplet spectral shape, that is a profile dominated by long-living collective modes arising from the quasi-conserved nature of the hydrodynamic variables: the densities of mass, momentum and energy.

It is reasonable to assume that the departure from the hydrodynamic limit at larger Q, ω values is smooth enough to be safely described by leaving the formal structure of the hydrodynamic equations unaltered, yet generalizing some transport and thermodynamic coefficients as suitable Q- and ω-dependent variables.

The opposite, $Q, \omega \to \infty$, regime customarily referred to as single-particle, or impulse approximation regime, can be easily understood by assimilating the fluid to an ensemble of rigid particles interacting via mutual collisions. In this extreme dynamic domain, the spectroscopic measurement probes such small volumes, $V \in (0, Q^{-3})$, and times, $t \in (0, \omega^{-1})$, that the only dynamic event it can detect is the free recoil of the single struck particle after the collision with the impinging photon (or neutron) beam and before its successive interaction with the first neighbouring atomic shell.

In this regime, the response of the system reflects the momentum distribution of the atom in its initial state, which, in the classical case, is ruled by the Boltzmann statistics. The corresponding spectrum is obtained by Fourier–Laplace transforming such a distribution, and it has a Gaussian profile centred at the energy of the struck atom's recoil.

Since no rigorous theory can predict the exact shape of the spectrum in the vast dynamic crossover between hydrodynamic and single-particle regimes, the measured or simulated spectral shapes are usually modelled using merely phenomenological extensions of the two limits to finite Q and ω's.

A possible unified theory should provide a consistent account of the gradual disappearance of collective modes dominating the spectral response at hydrodynamic scales and the parallel predominance of the ballistic, single-particle behavior. From the experimental side, the need for such a unified theory has become even more urgent after the development of IXS. In fact, the virtual lack of kinematic limitations of this method enables the experimental coverage of a relevant portion of the crossover between the hydrodynamic and the single-particle regimes by using a single spectrometer.

Although the use of memory function models demonstrated to be quite successful in describing the spectral evolution beyond the hydrodynamic regime, this approach suffers from inherent limitations. Specifically:

(1) First, it only enables a lineshape modelling which rests on a phenomenological *ansatz* on the time dependence of the memory function.
(2) In principle, a more accurate short-time description of the lineshape can be achieved by superimposing to the model the fulfillment of higher-order sum rules. However, this strategy loses value and physical ground when higher-order spectral moments with elusive meaning and complicated analytical structure are involved.
(3) Finally, "popular" models currently used for the memory function fail to predict the onset of additional inelastic features in the spectrum, whose emergence is, instead, frequently documented by both simulations and experiments (see, for instance, Chapter 8).

Some of these inherent drawbacks can be successfully overcome by using different approaches like the one based on the description of the lineshape as a sum of generalized Lorentzian terms (Bafile *et al.*, 2012). However, subsequent chapters will privilege memory function-based approaches — for historical reasons too — as these models are most frequently used to describe *IXS* spectra reported in the literature.

6.1. The two opposite regimes of the spectral shape

Before discussing various models describing the lineshape in the broad crossover region extending between hydrodynamic and single-particle regimes, it is useful to illustrate, as a reference, the spectral shape behaviour in these two opposite limits, where its analytical form is known rigorously.

6.1.1. The hydrodynamic regime (Berne and Pecora, 1976)

An analytical prediction of the hydrodynamic spectrum of a liquid was first proposed by Landau and Placzek in the early 1930s (Landau and Placzek, 1934), while a more recent derivation can be found in the textbook of Berne and Pecora (Berne and Pecora, 1976). The conceptual steps of this treatment are summarized below.

- The starting point is the formal expression of the five conservation laws of the (densities of) mass, three components of momentum and energy.
- The equations are written using spontaneous equilibrium fluctuations as dynamic variables and are linearized by retaining only the terms linear in these small fluctuations.[1]
- These fluctuations are represented in terms of thermodynamic functions' derivatives, and of both temperature (T) and pressure (P) fluctuations.
- The step above introduces in the hydrodynamic equations the fluctuations of two additional thermodynamic variables, i.e. T and P, increasing to seven the number of unknown fluctuating variables. Therefore, the solution of the hydrodynamic equations now requires the use of two additional constitutive equations: the Navier–Stokes equation for the viscosity and the Fourier law of the heat flow.
- This system of differential equations is then Fourier and Laplace transformed to be represented in the Q, ω domain.

---

[1]To get a general flavour of how this linearization works, one can consider, for instance, the fluctuation of the momentum density $(\rho_0 + \delta\varrho)\delta\mathbf{v}$, where $\rho_0 = nM$ is the equilibrium value of the mass density (the velocity $\mathbf{v}$ has a vanishing equilibrium value). The corresponding linearized fluctuation is given by the term $\rho_0\delta\mathbf{v}$ only. Similarly, when considering the fluctuation of the density of energy, the non-linear kinetic term $\rho_0 M(\delta\mathbf{v})^2$ is neglected as quadratic in the small fluctuation $\delta\mathbf{v}$, and so forth.

• At this stage, the correlation matrix $C(\omega)$ can be computed, and the density autocorrelation function can be derived from it by simple matrix inversion. Notice that, in this procedure, a significant simplification comes from the mutual orthogonality of most dynamic variables involved, i.e. from the circumstance that cross-correlations of these variables are vanishing.

As a result of the above treatment, the following analytical form is obtained for the linearized hydrodynamic spectrum of density fluctuations:

$$
\frac{S(Q,\omega)}{S(Q)} = \frac{A_h(Q)}{\pi} \frac{z_h(Q)}{\omega^2 + z_h^2(Q)}
$$

$$
+ \frac{A_s(Q)}{\pi} \left[\frac{z_s(Q)}{[\omega + \omega_s(Q)]^2 + z_s^2(Q)} + \frac{z_s(Q)}{[\omega - \omega_s(Q)]^2 + z_s^2(Q)} \right]
$$

$$
+ \frac{A_h(Q)}{\pi} b_s(Q) \left[\frac{[\omega + \omega_s(Q)]}{[\omega + \omega_s(Q)]^2 + z_s^2(Q)} \right.
$$

$$
\left. - \frac{[\omega - \omega_s(Q)]}{[\omega - \omega_s(Q)]^2 + z_s^2(Q)} \right], \tag{6.1}
$$

which is customarily referred to as the linearized hydrodynamics spectrum. Here $A_h(Q)$ and $A_s(Q)$ are Q-dependent area coefficients, and $b_s(Q) = [A_h(Q)z_h(Q)/(1 - A_h(Q)) + z_s(Q)]/\omega_s(Q)$, where $\omega_s(Q)$ and $z_s(Q)$ are, respectively, the shift and the half-width at half maximum (HWHM) of the (symmetric) inelastic modes, while z_h is the HWHM of the quasi-elastic one.[2]

The various shape parameters in Eq. (6.1) are involved functions of the hydrodynamic frequencies $c_s Q$, $D_T Q^2$ and ΓQ^2. Here $\Gamma = 1/2[(\gamma - 1)D_T/\gamma + \nu_{(L)}]$, where c_s, γ, D_T are the sound velocity, the constant-pressure-to-constant-volume specific heats ratio and the thermal diffusivity, respectively, while $\nu_L = \eta_L/\rho$ is the kinematic longitudinal viscosity, with $\rho = nM$ and η_L being, respectively, mass density and longitudinal viscosity. Over hydrodynamic scales,

---

[2]On a rigorous ground, the definition half width at half maximum of the Brillouin peaks is less straightforward as these individual peaks in the spectral profile also include terms with negative tails.

one has $\eta_L = \eta_b + 4/3\eta_s$, with η_s and η_b being the shear and the bulk components of the viscosity, respectively. Furthermore, $D_T = k/(\rho c_v)$, where k is the thermal conductivity and c_v the constant volume specific heat.

A complete expression of the linear hydrodynamic spectral shape as a function of these parameters can be found, for instance, in Chapter 10 of (Berne and Pecora, 1976). Here we are only interested in a suitable low Q approximation. Indeed, for most fluids, the condition $D_T Q^2, \Gamma Q^2 \ll c_s Q$ is fulfilled, and this provides the ground for a series expansion in the small variables $D_T Q^2$ and ΓQ^2 of the eigenvalue equation for the hydrodynamic matrix. Within a $O(Q^2)$ approximation, this expansion yields the following formulas for the spectral parameters of Eq. (6.1):

$$
\begin{cases}
\omega_s(Q) = c_s Q \\
z_s(Q) = \Gamma Q^2 = [(\gamma - 1)D_T + \nu_L]Q^2/2 \\
z_h(Q) = D_T Q^2 \\
A_s(Q) = \dfrac{1}{\gamma} \\
A_h(Q) = (\gamma - 1)/\gamma \\
b_s(Q) = b(Q) = Q(3\Gamma - \nu_L)/(\gamma c_s)
\end{cases}
\tag{6.2}
$$

Therefore, in this limit, the linearized hydrodynamic spectrum of Eq. (6.1) transforms to:

$$
\begin{aligned}
S_{RB}(Q,\omega) = \frac{S(0)}{\pi} &\left\{ \left(\frac{\gamma-1}{\gamma}\right) \frac{D_T Q^2}{\omega^2 + (D_T Q^2)} \right. \\
&+ \frac{1}{\gamma}\left[\frac{\Gamma Q^2}{(\omega + c_s Q)^2 + (\Gamma Q^2)^2} + \frac{\Gamma Q^2}{(\omega - c_s Q)^2 + (\Gamma Q^2)^2} \right] \\
&+ \frac{b(Q)}{\gamma}\left[\frac{\omega + c_s Q}{(\omega + c_s Q)^2 + (\Gamma Q^2)^2} \right. \\
&\left.\left. - \frac{\omega - c_s Q}{(\omega - c_s Q)^2 + (\Gamma Q^2)^2} \right]\right\}
\end{aligned}
\tag{6.3}
$$

The lineshape above is customarily referred to as the Rayleigh–Brillouin spectrum, and it consistently approximates the spectrum of density fluctuations in the extremely low Q, ω regime. Here the system appears as a continuous medium, whose response is probed over times much longer than any mesoscopic or microscopic event occurring in its interior.

The components of the Rayleigh–Brillouin spectrum are (from left to right in Eq. (6.3)):

(1) The central, or Rayleigh peak, described by the Lorentzian term having half-width $D_T Q^2$, which arises from heat fluctuations diffusing as a result of spontaneous temperature gradients in the fluid.

(2) The two Brillouin side peaks, i.e. the first term between square brackets. These peaks have positions $\pm c_s Q$ and half-width ΓQ^2 and relate to density fluctuations propagating in the form of acoustic waves.

(3) An additional spectral contribution (second term between square brackets) asymmetric around $\pm c_s Q$ and having negative high-frequency tails. This term sharpens the decay of the large ω tails of the Lorentzian modes, thus enabling the convergence of the second spectral moments.

As mentioned, the Rayleigh–Brillouin spectrum can be observed provided $D_T Q^2, \Gamma Q^2 \ll c_s Q$, i.e. if both the lifetime of the acoustic mode and the thermal diffusion time — $2\pi/(\Gamma Q^2)$ and $2\pi/(D_T Q^2)$ respectively — are much longer than the inverse of the acoustic frequency This implies, from one side, that thermal diffusion is perceived as extremely slow and, from the other, that acoustic excitations experience many oscillations in their lifetime. In other terms, the narrow widths of the peaks dominating $S_{RB}(Q, \omega)$ reflect the slowly decaying, or long-living, nature of density fluctuations in the hydrodynamic limit.

This is not surprising if one considers that the hydrodynamic equations involve a set of slowly changing variables, which become conserved (constants of motion) in the macroscopic limit.

An example of the Rayleigh–Brillouin triplet is provided in Figure 6-1.

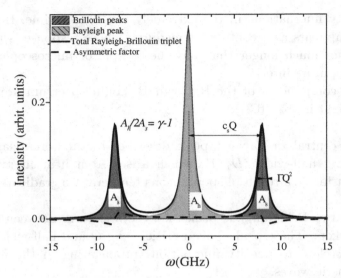

Figure 6-1: A typical Rayleigh–Brillouin triplet spectrum of a liquid (Eqs. 6.2–6.3). The individual contributions to the whole shape are represented by shaded areas and thick lines, as indicated in the legend.

Notice that the three modes of the Rayleigh–Brillouin triplet represent only a portion of the hydrodynamic modes, and specifically those coupling with density fluctuations. In fact, one might recognize that the conservation laws mentioned at the beginning of this section involve five constants of motions (the densities of mass, energy, and the three components of the momentum). Out of these five fluctuating variables, two do not couple with density fluctuations in the hydrodynamic, low Q and ω, limit; these are the modes involving the two shear or transverse components of the momentum. In the next section, we will consider a different representation of the hydrodynamic equations, which might help to clarify this point.

6.1.2. General considerations on the physical nature of hydrodynamic modes (Boon and Yip, 1980)

In principle, the actual choice of variables to represent the hydrodynamic equations is mostly left to the investigator's preference, although the statistical orthogonality (lack of correlation) of most of these variables is a useful asset.

As discussed in (Boon and Yip, 1980), it is particularly instructive to use as an alternative set of intensive fluctuating variables, the pressure, P, the density of entropy, S, and the longitudinal and transverse components of the current, i.e. $J_L = \rho_0 \nabla \cdot \mathbf{v}$ and $J_T^{y,z} = \rho_0 (\nabla \times \mathbf{v})^{y,z}$, respectively. Notice that longitudinal and transverse current fluctuations involve, respectively, velocity components either along or orthogonal to $\mathbf{Q}$, here assumed parallel to the x-axis.

To the $O(Q^2)$, the eigenvalue equation for the $z = i\omega$ Laplace transform of the hydrodynamic matrix yields the following roots:

$$z_\pm = \pm i c_s Q - \Gamma Q^2$$

$$z_0 = -D_T Q^2$$

$$z_{1,2} = - \left(\frac{\eta_s}{nM} \right) Q^2.$$

The first three eigenvalues are the same as the ones dominating the Rayleigh–Brillouin spectrum introduced in the last section.

The complex conjugate eigenvalues $z_\pm$ relate to pressure waves propagating at constant entropy (i.e. adiabatically) with a frequency $c_s Q$ and a lifetime $(\Gamma Q^2)^{-1}$. Indeed, in in thermally non-conductive fluids, acoustic propagation in the hydrodynamic limit takes place in adiabatic conditions, as the compression-decompression zones forming the acoustic wave oscillate more rapidly than thermal exchanges with the propagation environment.

Conversely, entropy fluctuations are essentially isobaric, that is they occur at constant pressure. The real eigenvalue z_0 is associated to a mode involving these constant-pressure entropy fluctuations, δS the circumstance that these fluctuations are associated with a real eigenvalue reveals their purely diffusive nature.

The $z_{1,2}$ eigenvalues define shear (transverse) density fluctuation modes, which have a merely diffusive character as indicated by their real-valued eigenvalue $-(\eta_s/nM)Q^2$, which is identical for the two modes due to the isotropy, i.e. the rotation invariance, of liquid systems.

From a physical point of view, the diffusive nature of shear modes mirrors the well-known inability of simple fluids to support shear propagation over long distances; this inability stems from the lack of rigidity, that is the weak shear restoring forces characteristic of fluid aggregates.

If one neglects the small coupling between pressure and entropy fluctuations — as reflected by the thermal contribution to the damping of the acoustic mode $\Gamma = [(\gamma - 1)D_T + \nu_L]Q^2/2$ — in the chosen representation, the hydrodynamic correlation matrix appears in the decoupled form schematized in Figure 6-2. In this diagonal matrix the two $z_\pm$ eigenmodes (acoustic or pressure modes) belong to the upper left minor, while the z_0 eigenmode belongs to the central one (entropy mode). These three longitudinal modes entirely define the spectrum of density fluctuations in the hydrodynamic limit, as transverse modes in this regime are uncoupled with density fluctuations.[3]

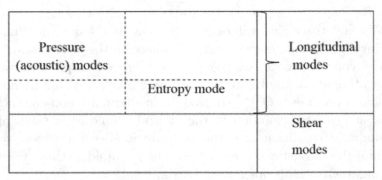

Figure 6-2: A schematic representation of the decoupled structure of the hydrodynamic matrix (see text).

---

[3]To understand this point, one might start from considering the Fourier transform of the microscopic density fluctuation $\delta n(\mathbf{Q}, t) = [dn(\mathbf{Q}, t)/dt]dt = \sum_{j=1}^{N} i\mathbf{Q} \cdot \mathbf{v}_j(t)dt \exp[i\mathbf{Q} \cdot \mathbf{R}_j(t)]$. The scalar product, represented by the "." symbol in the factor $i\mathbf{Q} \cdot \mathbf{v}(t)$, projects molecular velocities in the longitudinal direction, i.e. the one parallel to $\mathbf{Q}$; therefore, movements in the orthogonal plane have no effect on density fluctuations. This is true unless some microscopic interaction causes a coupling between transverse and longitudinal velocities. Similar considerations can be easily extended to the Fourier transform of hydrodynamic density fluctuations $\delta n_Q(t) = \int d\mathbf{r}\delta n(\mathbf{r}, t) \exp(i\mathbf{Q} \cdot \mathbf{r})$; in this case, it should be considered also that the lack of a large-scale order (symmetry) in the molecular arrangements prevents the coupling between transverse and longitudinal velocities over hydrodynamic lengthscales, where the matter appears as a continuum.

As discussed further below in this book, this scenario changes drastically at sufficiently large Q values, at which transverse and longitudinal excitations might become intertwined in some disordered systems, and the mere notion of mode polarization (either transverse or longitudinal) gradually loses physical significance.

Hydrodynamic spectrum and memory function

A straightforward derivation of the hydrodynamic spectrum can be achieved in the framework of the memory function formalism described in Chapter 5. To this purpose, it is worth noticing that the variable current become conserved upon approaching the macroscopic limit and its memory function features a Markovian, or instantaneous, component. This instantaneous relaxation is connected to all dissipative processes affecting the viscous flow which, in the hydrodynamic limit, appear as infinitely rapid.

Furthermore, a consistent account of the slower dissipation processes involving thermal fluctuations is provided by an additional exponential term, customarily referred to as the Rayleigh contribution, as it corresponds to the homonymous peak in the hydrodynamic spectrum. Overall, the hydrodynamic form of the memory function reads as:

$$m_L(Q \to 0, t) = \omega_0^2(\gamma - 1)\exp(-\gamma D_T Q^2 t) + 2\nu_L Q^2 \delta(t), \qquad (6.4)$$

where $\omega_0 = c_T Q$, where $c_T = k_B T/[MS(0)]$ is the isothermal sound velocity.

Upon inserting real and imaginary parts of the Laplace transform, $\tilde{m}_L(Q \to 0, z = i\omega)$, of Eq. (6.4) into the expression of $S(Q, \omega)$ in Eq. (5.29), one eventually retrieves the Rayleigh–Brillouin triplet.

In summary, the hydrodynamic memory function in Eq. (6.4) consists of a first exponential term describing the Rayleigh or thermal contribution, while the second term, $\propto \delta(t)$, accounts for all viscous rearrangements that in the hydrodynamic limit appear infinitely rapid.

As discussed further below, in many fluids, the Rayleigh contribution is extremely slow and well approximated by a constant term within the typical time intervals probed by a spectroscopic measurement exploring the hydrodynamic regime. This yields a nearly flat contribution to the memory function decay and an extremely narrow central peak in the spectrum.

6.1.3. *The single-particle regime*

Let us now make an enormous leap in the (Q, ω)-plane and give a closer look at the opposite limit, *i.e.* the single-particle one. Over the short times and distances characteristic of this limit, the probed dynamic event reduces to the free recoil of the single struck atom after the collision with the probe and before any further interaction with the surrounding atomic cage. Under these conditions, the free streaming of the struck atom can be described by the equation for the ballistic motion of a free particle.

The interpretation of this dynamic behaviour is particularly straightforward for particles resembling rigid spheres, as they interact through instantaneous "contact" events (collisions). In a real fluid, atoms remain continuously connected with neighbours by finite-distance forces, which, in principle, prevent their recoil from being entirely free. Nonetheless, if the energy transferred in the scattering event largely exceeds any local interaction strength, the struck atom, is freed for short times from all connections with its environment. Therefore, in this so-called impulse approximation (IA) regime, it is safe to assume that no external force acts on the isolated system formed by the photon and the target atom. Here the response of the struck atom reflects the distribution of momenta in its initial state, each of these momenta yielding a sizable contribution to the scattering intensity.

Upon approaching the asymptotic IA limit, correlations between distinct atoms, *i.e.* collective effects in the dynamic response, gradually vanish, as probed distances are here much smaller than first neighbouring atoms' separations. In this high Q regime, an incoherent approximation of the lineshape holds validity. Within this approximation, the correlation involving pairs of distinct atoms vanishes and the response of the system is uniquely determined by self-correlations.[4]

Consequently, in this so-called incoherent approximation range, the "coherent" Q-oscillations of $S(Q)$ are damped off, and this variable closely approaches its asymptotic unit value (see Figure 6-3).

Consistently, the dynamic response in this regime can be described by the self-component of the van Hove function in Eq. (2.79) or, in the reciprocal space, by its Fourier transform $F_s(Q,t)$.

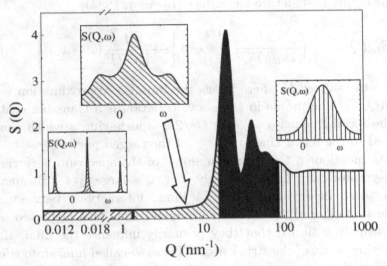

Figure 6-3:   Sketch of the expected evolution of the spectral shape from a liquid sample across various Q windows characterized by different behaviours of the $S(Q)$ profile, plotted in semi-logarithmic scale. Courtesy of F. Bencivenga.

---

[4]Notice that in this limit the intermediate scattering function introduced in Chapter 2 reduces to its *self* part, $F_s(Q,t) = 1/N \sum_{j-1}^{N} \exp[i\mathbf{Q} \cdot \mathbf{R}_j(t)] \exp[-i\mathbf{Q} \cdot \mathbf{R}_j(0)]$, which only involves the positions of the jth atom at different times. However, the other $N-1$ atoms of the system still influence the value of this self-correlation, as they determine the time-evolution of the jth atom's position, $\mathbf{r}_j(t)$.

After being hit by the incident beam, the target atom re-radiate the photons, behaving as a moving point-source of scattering. The intensity it emits is thus Doppler-shifted by an amount which depends on the momentum of the struck atom in its initial state. When the response is averaged over all atoms in the illuminated portion of the sample, the resulting spectral density becomes simply related to the momentum distribution of the atoms in their initial states, which, for classical monatomic systems, is determined by the Gaussian Maxwell–Boltzmann factor $\exp(-p^2/2Mk_BT)$. The Fourier transform of such momentum distribution leads to the following Gaussian shape of the dynamic structure factor (see (Lovesey, 1984)):

$$S_{IA}(Q,\omega) = \left(\frac{M}{2k_B\pi TQ^2} \right)^{1/2} \exp\left[-\frac{M}{2k_BTQ^2}(\omega - \omega_r)^2 \right], \quad (6.5)$$

where the suffix "*IA*" here labels the Impulse Approximation value of $S(Q,\omega)$. The profile in the above equation is a Gaussian centred at the recoil frequency $\omega_r = \hbar^2Q^2/2M$ and having a width simply related to the mean kinetic energy of the tagged particle.

As mentioned, the Gaussian shape of the spectrum is retrieved at asymptotically large Q's only. For Q's large, yet still smaller than those belonging to the *IA* limit, interactions between the single particle and the local environment cannot be neglected, and it is usually assumed that they primarily influence the final, after-scattering, state of the struck atom. These so-called final state effects manifest themselves through small deviations of the spectral tails from their Gaussian limiting shape.

The gradual transformation of low Q collective modes into these final state spectral features is a topic of prominent interest still in search of a unified and firm theoretical understanding. However, the expected qualitative aspects of this wide dynamic crossover are schematically illustrated in Figure 6-3. At small Q's, the spectral shape rapidly evolves from the Rayleigh–Brillouin profile toward a more complex triplet structure, in which the dominant modes become increasingly damped. In particular, when the momentum transfer is comparable with the inverse of the interparticle separation, *i.e.* in the region approaching the main $S(Q)$ peak, one probes the so-called "mesoscopic regime", roughly indicated by the arrow, where the three main spectral features become much broader. A shadowed band approximately locates the so-called kinetic regime, where the lineshape has nearly lost any neat signature of inelastic modes. The Gaussian profile centred at the recoil frequency is retrieved asymptotically only, i.e. when the Q-oscillations of $S(Q)$ damp off and the latter reaches the limiting unit value.

6.2. Modelling the lineshape at the departure from the hydrodynamic limit

As mentioned in the introductory part of this chapter, in the absence of a firm theoretical prediction, the departure of the spectral shape from the hydrodynamic Rayleigh–Brillouin triplet is usually described by best-fitting measured or simulated lineshapes with suitable models. When performing this best-fit procedure, it is advisable to impose some physical constraints to the Q-dependence of the shape parameters in the model. These constraints are needed to reduce the number of free parameters in the fitting routine and to confer a more solid physical ground to the lineshape description. For instance, one can superimpose the sum rules fulfilment or adopt reasonable approximations on the thermodynamic and transport properties as suited to the sample at hand and the specific thermodynamic conditions investigated.

Also, to corroborate the best-fitting results, it is useful to compare them with the outcome of complementary measurements or computations, either performed in joint experiments, or available from the literature. To this purpose, the Q-dependent best-fitting parameters derived from *IXS* spectra are often extrapolated at $Q = 0$ to be compared with low Q, ω spectroscopic measurements. In other cases, simulation or experimental results can be considered for direct comparison or to be incorporated in the model to fix its spectral moments or other shape parameters.

Regardless of the adopted best-fit and data analysis strategy, some knowledge of model options available is essential. Further below, we provide a succinct overview of some popular models used to best-fit *IXS* and *INS* spectra at the departure from the hydrodynamic regime, or in more remote dynamic domains.

It is essential to recognize that, well-beyond the hydrodynamic limit, transport coefficients become local, rather than global, properties of the system; as mentioned, this circumstance mirrors the gradual loss of homogeneity and stationarity of matter upon approaching mesoscopic scales. Spatial inhomogeneities might arise either from the atomistic nature of matter, or, at longer distances, from the possible existence of a mesoscale arrangement, such as clusters, heterostructures, and so forth. Conversely, losses of stationarity can be brought about by the coupling of density fluctuations with internal degrees of freedom of the fluid, giving origin to the relaxation phenomena extensively discussed in Chapter 8. Clearly, in the reciprocal space, the time-dependence translates into an ω-dependence of model parameters, while space inhomogeneities are while reflected by a Q-dependence.

Within the Mori–Zwanzig formalism introduced in Chapter 5, the ω-generalization of the transport coefficients is accounted for by using a suitable model for the time-decay of the memory function, which contains some parameters having an unknown Q-dependence.

A detailed account of various microscopic processes contributing to the time-decay of the memory function would in principle entail the use of many timescales — or even a continuous distribution of them; however, a model containing only a few dominant relaxation times, or even a single one, is often a reasonable approximation advised by conventional wisdom. In fact, an excessive number of free model parameters would ultimately jeopardize the reliability of the lineshape analysis outcome.

For these reasons, the most popular models of the memory function succinctly discussed further below are based on simple approximation entailing a minimum number of free parameters. Typically used approximations contains suitable linear combinations of a δ-function, an exponential, and a constant term.

6.2.1. *Generalized Hydrodynamics models*

The so-called Generalized Hydrodynamics (*GH*) models rely on the assumption that the departure from the Rayleigh–Brillouin shape is smooth enough to be safely described by leaving the formal structure of the hydrodynamic spectrum unaltered, yet making allowance for an (unknown) Q-dependence of relevant shape parameters.

The analytical form of the spectrum is analogous to the linearized hydrodynamic spectrum in Eq. (6.1), here rewritten in a more compact form for convenience:

$$
\frac{S(Q,\omega)}{S(Q)} = \frac{A_h(Q)}{\pi}\frac{z_h(Q)}{\omega^2 + z_h^2(Q)}
$$

$$
+ \frac{A_s(Q)}{\pi}\left[\frac{z_s(Q) + b_s(Q)[\omega + \omega_s(Q)]}{[\omega + \omega_s(Q)]^2 + z_s^2(Q)}\right.
$$

$$
\left. + \frac{z_s(Q) - b_s(Q)[\omega - \omega_s(Q)]}{[\omega - \omega_s(Q)]^2 + z_s^2(Q)}\right]; \tag{6.6}
$$

all shape parameters here have an unknown dependence on Q, which is usually determined *a posteriori* only from the best-fitting of the model to the spectra measured at various Q values. The lineshape expressed by the above equation defines the family of the Generalized Hydrodynamics, or *GH*, models. Indeed, various *GH* models might differ in specific constraints imposed on the Q-dependence of shape parameters.

In Appendix 7A it will be shown that a similar model profile can also be derived from a kinetic theory approach by expressing the spectral shape as the sum of generalized (complex) Lorentzian terms.

Within the Mori–Zwanzig formalism, a *GH* model of the lineshape can be generated using a trivial finite-Q generalization of the simple hydrodynamic second-order memory function in Eq. (6.4), given as follows:

$$
m_L(Q,t) = \omega_0^2(Q)\left[\gamma(Q) - 1\right]\exp\left[-\gamma(Q)D_T(Q)Q^2 t\right]
$$

$$
+ 2\nu_L(Q)Q^2\delta(t). \tag{6.7}
$$

6.2.2. *Single timescale approximation of the memory decay, or pure viscoelastic model*

While *GH* models rely on the assumption that the departure from the hydrodynamic regime solely influences the *Q*-dependence of transport parameters, the so-called Molecular Hydrodynamics (*MH*) models explicitly tackle the frequency- or time-dependence of the memory function. The simplest model belonging to this category is customarily referred to as simple viscoelastic or Debye model, and rests on the assumption that $m_L(Q,t)$ has the form of a single exponential law:

$$m_L(Q,t) = \Delta_2(Q)\exp\left[-t/\tau(Q)\right], \tag{6.8}$$

where $\Delta_2(Q)$ is the relaxation strength. From the sum rule fulfillment, it follows that $\Delta_2(Q)$ coincides with the parameter Δ_2 in Eq. (5.26). The timescale $\tau(Q)$ is the finite-Q generalization of the Maxwell relaxation time, as discussed further below. Although the simple viscoelastic model in many situations represents a clear oversimplification, it was found adequate to describe terahertz relaxation phenomena in various simple systems (Lewis and Lovesey, 1977).

Simple viscoelastic model and hydrodynamic limit

Equation (6.8) can keep consistency with the hydrodynamic memory function (Eq. (6.4)) under the following conditions:

(1) At very low Q's, $\tau(Q)$ becomes much smaller than the typical timescale defining the decay of density fluctuations, i.e. in this limit, the viscoelastic memory reduces to a Markovian form. Namely: $\lim_{Q\to 0} m_L(Q,t) \propto \delta(t)$.

(2) The thermal decay of the memory function (i.e. the exponential, Rayleigh, in Eq. (6.4)) has negligible strength, as follows from the condition $\gamma(Q) \approx 1$, which is an accurate approximation for many liquid metals.

Let us now have a closer look at the macroscopic, or quasi-macroscopic, limit, where the Maxwell description of viscoelasticity applies. Within the Maxwell model, the velocity of an acoustic wave depends on its frequency. More specifically, upon frequency increase, the sound velocity evolves from its viscous value, c_0, to the elastic one, c_∞. The frequency demarcating the transition from the viscous (or liquid-like) behavior to the elastic (or solid-like) one is equal to $1/\tau$, where τ is the (Maxwell) relaxation time. Within the Maxwell's model, the (longitudinal) viscosity is equal to:

$$\nu_L = \tau(c_\infty^2 - c_0^2).$$

If one neglects the Rayleigh or thermal contribution to the memory function, as assumed in Eq. (6.8), and considers the points (1) and (2) above, the consistency of Eq. (6.8) with the hydrodynamic memory in Eq. (6.4) (without the Rayleigh term), implies the equality of the two integrals:

$$\lim_{Q\to 0} \int_0^\infty dt \Delta_2(Q) \exp\left[-t/\tau(Q)\right] = \int_0^\infty dt \left(2\nu_L Q^2\right) \delta(t).$$

The two integrals in the equation above can be simply evaluated by using $\int_0^\infty dt\, \delta(t) = 1/2$. Furthermore, considering the Maxwell expression of ν_L reported above, and that $\tau(Q=0) = \tau$, one has $\Delta_2(Q \to 0) = (c_\infty^2 - c_0^2)Q^2$. This relationship can be naturally extended to finite Q values, which yields:

$$\Delta_2(Q) = \left(c_\infty^2(Q) - c_0^2(Q)\right) Q^2,$$

where the involved parameters are finite Q generalization of the corresponding parameters appearing in the Maxwell theory.

Notice that, for most liquids, the relaxation time so evaluated remains finite and typically spans the picosecond range. Conversely, owing to the conservation laws valid at macroscopic distances (see Chapter 5), the density tends to become a constant of motion at extremely low Q's. This means that equilibration time of density fluctuations diverge, and, at some finite Q, it largely exceeds $\tau(Q)$, which justifies the Markovian approximation in the hydrodynamic expression of the memory function in Eq. (6.14).

The superimposition of the first three sum rules to the viscoelastic model in Eq. (6.8) leads to:

$$c_o^2(Q) = \frac{1}{Q^2} \frac{\langle \omega^2 \rangle}{\langle \omega^0 \rangle} = \frac{k_B T}{M S(Q)} \tag{6.9}$$

$$c_\infty^2(Q) = \frac{1}{Q^2} \frac{\langle \omega^4 \rangle}{\langle \omega^2 \rangle} = \frac{1}{Q^2} \left(3 \frac{k_B T}{M} Q^2 + \Omega_0^2 - \Omega_Q^2 \right) \tag{6.10}$$

$$\Delta_2(Q) = \left[c_\infty^2(Q) - c_0^2(Q) \right] Q^2 = \frac{\langle \omega^4 \rangle}{\langle \omega^2 \rangle} - \frac{\langle \omega^2 \rangle}{S(Q)}, \tag{6.11}$$

where $S(Q) = \langle \omega^0 \rangle$, and we have introduced the parameter:

$$\Omega_Q = \frac{n}{M} \int d\mathbf{r} g(r) \exp(-iQx) \, \nabla_x^2 V(r), \tag{6.12}$$

where we have assumed that the x-axis is parallel to $\mathbf{Q}$. At low Q values, the parameter defined above approaches the Einstein frequency Ω_0, previously introduced in Chapter 2 (see Eq. 2.21).

Equation (6.9) can be seen as the finite Q extension of the macroscopic ($Q = 0$) compressibility theorem (Egelstaff, 1967), leading to expresss the isothermal sound velocity as $c_T = \sqrt{k_B T / [M S(0)]}$. In fact, within the validity of the pure viscoelastic model, $\gamma(Q) = 1$, hence, $c_T(Q) = c_s(Q)/\gamma(Q) = c_s(Q)$.

As mentioned, the viscoelastic model hinges on the assumption that, upon frequency increase, the response of a fluid changes from liquid-like (viscous) to solid-like (elastic). Such a transformation affects the sound velocity as follows:

(1) In the viscous, low frequency regime, the sound velocity is proportional to $\sqrt{B(Q)/\rho}$, with $B(Q)$ being the generalized bulk modulus as appropriate for a liquid.
(2) In the elastic regime, it becomes proportional to $\sqrt{(B(Q) + 4G(Q)/3)/\rho}$, with $G(Q)$ being the generalized shear modulus, as adequate for solids. Notice that, being $G(Q) > 0$, the sound velocity increases across the viscous-to-elastic transition.

It appears that the imposition of the sum rules (Eqs. (6.9)–(6.11)) ultimately sets a link between the generalized moduli and the spectral moments which confers to the model a deep-seated physical ground. This also emphasizes some experimental challenges to be faced when detecting viscoelastic effects in the measured lineshape. In fact, in the viscous regime the acoustic frequency is equal to $\sqrt{\langle\omega^2\rangle/\langle\omega^0\rangle}$, while it becomes equal to $\sqrt{\langle\omega^4\rangle/\langle\omega^2\rangle}$ in the elastic one; this implies that a reliable experimental detection of non-viscous (i.e. solid-like, or elastic) effects in the spectrum ultimately rests on a measurement of spectral tails precise enough to enable a firm determination of the first three even spectral moments. This requirement appears particularly challenging, especially for spectroscopic techniques such as *IXS*, for which the high-frequency intensity is dominated by the slowly decaying resolution tails.

From a complementary perspective, one can consider a fourth moment-truncated short-time expansion of the density autocorrelation function:

$$\langle\delta n(Q,t)\delta n^*(Q,0)\rangle \approx \sum_{n=0}^{2}(-1)^n\langle\omega^{2n}\rangle\frac{t^{2n}}{2n!},$$

which consistently describes the early-stage response of the system up to the $O(t^4)$ infinitesimal.

For instance, solid-like effects are not accounted for in the Rayleigh–Brillouin triplet of Eq. (6.3) which has converging moments up to the second moment only, thus providing a description of the short-time dynamics consistent to the $O(t^2)$ order only. Indeed, when probing the dynamics of a simple fluid in the hydrodynamic regime, elastic effects cannot be properly observed, and the system's response exhibits its fully relaxed, or viscous, behavior.

6.2.3. *Molecular Hydrodynamics models*

The simple viscoelastic model defined above provides the simplest example of the category of Molecular Hydrodynamics (*MH*) models; these models account for non-hydrodynamic effects by suitable modifications of the time-decay of $m_L(Q,t)$, which includes non-Markovian terms arising from the onset of finite timescale relaxation processes. Very often, such a time-decay can be consistently approximated by the following general form:

$$m_L(Q,t) = \omega_0^2(Q)\left[\gamma(Q) - 1\right]\exp\left[-\gamma(Q)D_T(Q)Q^2t\right] + K_L(Q,t)Q^2.$$

The requirement of consistency with the $Q = 0$ hydrodynamic result in Eq. (6.4) imposes:

$$2\nu_L\delta(t) = \lim_{Q\to 0} K_L(Q,t). \qquad (6.13)$$

The above equation can be integrated over time to obtain:

$$
\begin{aligned}
\nu_L &= \lim_{Q\to 0} \int_0^\infty dt\, K_L(Q,t) \\
&= \lim_{Q\to 0} \int_0^\infty dt \int_V d\mathbf{r}\, K_L(\mathbf{r},t)\exp(i\mathbf{Q}\cdot\mathbf{r}) \\
&= \int_0^\infty dt \int_V d\mathbf{r}\, K_L(\mathbf{r},t). \qquad (6.14)
\end{aligned}
$$

Notice that, while performing the integration, we used that $\int_0^\infty dt\delta(t) = 1/2 \int_{-\infty}^\infty dt\delta(t) = 1/2$. From the comparison of the two extremes of the above chain of identities, one can deduce that the longitudinal viscosity, ν_L, is a time and space average of the local variable $K_L(\mathbf{r},t)$.

In summary, *MH* models of the spectral lineshape can be generated by using the following form for the second-order memory function:

$$
\begin{aligned}
m_L(Q,t) &= \omega_0^2(Q)\left[\gamma(Q) - 1\right]\exp\left[-\gamma(Q)D_T(Q)Q^2t\right] \\
&\quad + 2\nu_L(Q,t)Q^2. \qquad (6.15)
\end{aligned}
$$

Notice that this expression becomes identical to the *GH* expression in Eq. (6.7) if the term $\nu_L\delta(t)$ is replaced by the generalized longitudinal viscosity $\nu_L(Q,t) \equiv K_L(Q,t)$.

The *MH* model defined above is the sum of two terms:

The generalized Rayleigh (thermal) contribution

The first term on the right-hand side of Eq. (6.15) is the finite Q's extension of the hydrodynamic Rayleigh mode arising from thermal diffusion and responsible for the central peak in the Brillouin triplet.

The generalized acoustic damping

The second term on the right-hand side of Eq. (6.15), proportional to the longitudinal kinematic viscosity $\nu_L(Q, t)$, accounts in principle for all relaxation processes affecting the viscosity and determining the damping of propagating acoustic waves.

To become suitable for practical use, the memory function in Eq. (6.15) still needs some *ansatz*, which makes the time-dependence of $\nu_L(Q, t)$ explicit.

A broadly used recipe prescribes the following time-dependence:

$$\nu_L(Q, t) = \Delta_2(Q)\exp\left[-t/\tau(Q)\right] + 2\Gamma_0(Q)\delta(t). \qquad (6.16)$$

Again, the consistency with the hydrodynamic results imposes that all Q-dependent coefficients join their macroscopic counterparts in the $Q \to 0$ limit.

Overall, the time-decay of the generalized kinematic longitudinal viscosity defined above is the sum of two terms:

The viscoelastic contribution

The first exponential term in Eq. (6.16) accounts for the evolution, upon time increase, from the solid-like to the liquid-like regimes, in which density fluctuations are respectively unrelaxed or fully relaxed. This crossover is usually referred to as "viscoelastic" since "viscous" (liquid-like), and "elastic" (solid-like) aspects coexist in it, as previously discussed in Section 6.2.2 and elucidated in further detail, in Chapter 8.

The instantaneous contribution

The Markovian contribution to Eq. (6.16), *i.e.* the term $2\Gamma_0(Q)\delta(t)$, describes the coupling of density fluctuations with the ultrafast intramolecular dynamics. Experimental and computational evidence suggest that the amplitude of this term exhibits a quadratic Q-trend and a weak dependence on thermodynamic conditions (Ruocco, Sette *et al.*, 2000).

An analysis of *IXS* spectra based on an MH modeling has been successfully carried out in a variety of samples definitely broader than the one strictly considered in this book and including noble gases (Cunsolo, Pratesi *et al.*, 1998; Cunsolo, Pratesi *et al.*, 2000; Verbeni, Cunsolo *et al.*, 2001; Bencivenga, Cunsolo *et al.*, 2009), diatomic liquids (Bencivenga, Cunsolo *et al.*, 2006; Bencivenga, Cunsolo *et al.* 2007), liquid metals (Scopigno, Balucani *et al.*, 2000; Scopigno, Balucani *et al.*, 2002; Monaco, Scopigno *et al.* 2004; Cazzato, Scopigno *et al.*, 2008; Hosokawa, Pilgrim *et al.*, 2008; Scopigno, Balucani *et al.*, 2009), hydrogen-bonded systems (Cunsolo, Ruocco *et al.*, 1999; Monaco, Cunsolo *et al.*, 1999; Angelini, Giura *et al.*, 2002), and glass formers (Fioretto, Buchenau *et al.*, 1999; Comez, Fioretto *et al.*, 2002; Scarponi, Comez *et al.*, 2004).

6.2.4. *Viscoelasticity and generalized transport parameters*

As mentioned, the simple viscoelastic model rests on the often oversimplified assumption of a negligible thermal contribution to the memory function decay. While making use of this simple approximation, we now discuss the physical meaning of real and imaginary parts of the complex variable $m_L(Q, \omega)$.

As the contribution from thermal fluctuation is discarded, the latter variable can be directly connected to real and imaginary parts of the generalized (complex) longitudinal kinematic viscosity $\tilde{\nu}_L(Q, z = i\omega)$, through:

$$\tilde{m}_L(Q, z = i\omega) \approx [\nu_L'(Q,\omega) - i\nu_L''(Q,\omega)]Q^2. \tag{6.17}$$

To gain some insight into the physical meaning of real and imaginary parts in the above equation, it is useful to compare the general form of $S(Q, \omega)$ with the Rayleigh–Brillouin triplet without the contribution of thermal diffusion. To perform this comparison, one can plug

Eq. (6.17) into the formula linking $S(Q,\omega)$ to the imaginary and real components of the memory function (Eq. (5.29)) and compare it with the one obtained using the hydrodynamic memory function in Eq. (6.4) (without thermal contribution); more easily the same result can be obtained by inserting $\gamma =1$ in the hydrodynamic expression (Eq. 6A.2 in Appendix 6A). The corresponding profiles are reported below for convenience:

$$
\begin{cases}
General : \dfrac{S(Q,\omega)}{S(Q)} \\
\qquad = \dfrac{1}{\pi} \left\{ \dfrac{\omega_0^2(Q)\nu_L'(Q,\omega)Q^2}{[\omega^2 - \omega_0^2(Q) - \omega\nu_L''(Q,\omega)Q^2]^2 + \omega^2[\nu_L'(Q,\omega)Q^2]^2} \right\} \\
Hydrodynamic : \dfrac{S(Q,\omega)}{S(Q)} \bigg|_{Q,\omega \to 0} \\
\qquad = \dfrac{1}{\pi} \left\{ \dfrac{\omega_0^2(Q \to 0)\nu_L Q^2}{[\omega^2 - (c_s Q)^2]^2 + \omega^2(\nu_L Q^2)^2} \right\}.
\end{cases}
$$

Here $\omega_0^2(Q \to 0) = c_T Q = c_T Q$. One can readily appreciate the overall "similarity" between the two spectral profiles reported above. In particular, the consistency requirement between the two expression above lends itself to the following finite Q- and ω-generalizations of the sound velocity:

$$c_s \to c_s(Q,\omega) = \frac{1}{Q}\sqrt{\omega_0^2(Q) + \omega\nu_L''(Q,\omega)Q^2}, \qquad (6.18)$$

and the viscous damping:

$$\nu_L Q^2 \to \nu_L''(Q,\omega)Q^2. \qquad (6.19)$$

Notice that the presence of frequency-dependent "shape parameters" changes the form of the spectrum; among these changes, one of the most evident is the emergence of additional relaxational modes in the spectrum, which are centred at $\omega = 0$, as discussed in the next chapter.

In summary, the real part of the generalized viscosity accounts for viscous dissipation processes affecting acoustic propagation, while the imaginary part is the one responsible for the viscoelastic increase of the sound velocity. The expression of $\tilde{\nu}_L(Q, z = i\omega)$ can be easily worked out for the pure viscoelastic model. Explicitly:

$$\tilde{\nu}_L(Q, z = i\omega) =$$

$$= (\Delta_2(Q)/Q^2) \left\{ \int_0^\infty dt \, \exp\left[\frac{-t}{\tau(Q)}\right] cos(\omega t) \right.$$

$$\left. - i \int_0^\infty dt \, \exp\left(\frac{-t}{\tau(Q)}\right) sin(\omega t) \right\}.$$

After performing the integrals above, the imaginary and real parts of the viscosity assume the following respective expressions:

$$\nu'_L(Q, \omega) = \frac{1}{\tau(Q)} \frac{c_\infty^2(Q) - c_0^2(Q)}{\omega^2 + [1/\tau(Q)]^2} \qquad (6.20)$$

$$\nu''_L(Q, \omega) = \omega \frac{c_\infty^2(Q) - c_0^2(Q)}{\omega^2 + [1/\tau(Q)]^2}. \qquad (6.21)$$

Upon inserting Eq. (6.21) into Eq. (6.18), while using $\omega_0(Q) = c_s(Q)Q$, one obtains the following expression for the generalized sound velocity:

$$c_s(Q, \omega) = \left[c_0^2(Q) + \omega^2 \frac{c_\infty^2(Q) - c_0^2(Q)}{\omega^2 + [1/\tau(Q)]^2} \right]^{\frac{1}{2}},$$

which defines the frequency dependence of $c_s(Q, \omega)$. It appears that, in the two opposite — zero and infinite frequency — limits, $c_s(Q, \omega)$ tends to $c_0(Q)$ and $c_\infty(Q)$, respectively.

The equation above is a finite-Q generalization of the formula defining the viscoelastic transition of the sound velocity occurring at $Q = 0$ in a viscoelastic fluid (Herzfeld and Litovitz, 1965). The ω-dependence is characterized by an inflexion point at $1/\tau(Q)$, which is usually referred to as the centre of the viscoelastic crossover.

Also, the generalized damping can be simply derived by inserting Eq. (6.20) into Eq. (6.19), which yields:

$$\nu'(Q, \omega)Q^2 = \frac{Q^2}{\tau(Q)} \frac{c_\infty^2(Q) - c_0^2(Q)}{\omega^2 + [1/\tau(Q)]^2}.$$

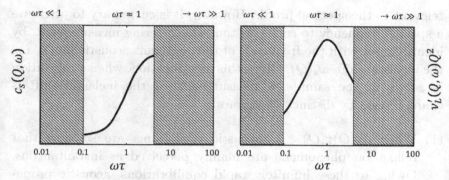

Figure 6-4: The frequency dependence of the generalized sound velocity and damping at a given Q value (left and right plot respectively, see text).

This ω-dependence has a form of a Lorentzian centered in $\omega = 0$, having halfwidth $1/\tau(Q)$ and amplitude $\tau(Q)\left[c_\infty^2(Q) - c_0^2(Q)\right]Q^2$. The frequency dependencies of the generalized sound velocity and acoustic damping are schematically displayed in Figure 6-4, as obtained for a generic fixed Q value.

We conclude this section with some general considerations on the link between viscoelasticity and relaxation phenomena as it emerges from this and previous sections.

The presence of a spontaneous, or scattering-generated acoustic wave can be seen as a time-dependent perturbation of the fluid. The fluid responds to such a perturbation through some dissipative channels which cause an energy redistribution affecting the acoustic wave in favour of some internal degrees of freedom of the system, having a characteristic timescale $\tau(Q)$. These energy rearrangements ultimately drive the system to recover to a new local equilibrium within a (relaxation) time comparable to $\tau(Q)$. The mere existence of such a "privileged" timescale lends itself to identify two different regimes of the acoustic propagation, the viscous and the elastic, which depend on how the probed frequency ω compares with $1/\tau(Q)$.

At this stage, an important question is how this ω-dependence manifests itself in a scattering experiment, typically characterizing the Q- rather than the ω-dependence of transport parameters relevant to the spectral profile. However, it is customary to attribute a specific frequency to each constant Q scattering measurement, by identifying it with the frequency of the dominant acoustic mode, *i.e.* the inelastic shift, $\omega_s(Q)$. Given this identification, when a relaxation is active in the sample, a measurement of the inelastic shift is characterized by distinct Q-regimes:

(1) When $\omega_s(Q)\tau(Q) \ll 1$, acoustic oscillations are so slow that relaxation phenomena are mainly perceived as instantaneous. Owing to these infinitely rapid equilibrations, acoustic propagation takes place over successive local equilibria of the system (**viscous limit**).

(2) Conversely, when $\omega_s(Q)\tau(Q) \gg 1$, internal degrees of freedom of the system are too slow to efficiently dissipate the energy carried by the acoustic wave, which mainly propagates without energy losses, *i.e.* elastically (**elastic limit**).

For intermediate acoustic frequencies $(\omega_s(Q)\tau(Q) \approx 1)$, the response of the fluid embodies both viscous and elastic aspects, thus being usually referred to as "viscoelastic".

Due to the progressive decrease of acoustic dissipation, the energy of the acoustic wave increases systematically across the viscous-to-elastic crossover and so does the sound velocity; the viscosity follows instead the opposite trend, which reflects the gradual freezing of dissipative processes toward the elastic regime.

6.3. Approximating the measured spectral shape: few general and practical issues

When modelling the spectral shape measured in a spectroscopic experiment, the signal is usually approximated by a model profile, $I_M(Q,\omega)$, having the following general form:

$$I_M(Q,\omega) = A\left[I(Q,\omega) \otimes R(\omega)\right] + B, \qquad (6.22)$$

where $R(\omega)$ is the energy resolution function, A is an overall intensity factor, while B accounts in principle for both the spectral background and the dark counts of the detector; sometimes the latter term, rather than being constant, is assumed to depend linearly on ω.

As mentioned in the introduction of this chapter, the modelling of the lineshape is almost entirely classical. In practice, quantum deviations are assumed to influence only the statistical population of the generic state of the system associated to an energy $\hbar\omega$, via the detailed balance principle. This principle states that, at equilibrium, each excitation process should be accompanied by the reverse one, as the probability of the transition to a given excited state cannot depend on how many atoms are gathered in such an excited state.

It is reasonable to assume that the number of excited $\hbar\omega$-states of the system at equilibrium, $n(\hbar\omega)$, is determined by the Bose statistics, according to which $n(\hbar\omega) = [\exp(\hbar\omega/k_BT) - 1]^{-1}$. The balance between the number of $\hbar\omega$ and $-\hbar\omega$ excitations at equilibrium implies that:

$$\frac{n(\hbar\omega)}{n(-\hbar\omega)} = \frac{I(Q,\omega)}{I(Q,-\omega)} = \exp\left(-\frac{\hbar\omega}{k_BT}\right).$$

The above condition can be fulfilled by the spectral shape in infinite possible ways, and the one considered in all IXS works is the multiplication of the classical profile $S(Q,\omega)$ by the frequency-dependent pre-factor $\hbar\omega/k_BT[n(\hbar\omega) + 1]$. Explicitly:

$$I(Q,\omega) = \frac{\hbar\omega}{k_BT}\left[\frac{1}{1 - \exp(-\hbar\omega/k_BT)}\right] S_M(Q,\omega), \quad (6.23)$$

where $S_M(Q,\omega)$ is the analytical profile approximating the "classical" or symmetric part of the spectrum, which is the one described by all lineshape models discussed in previous sections.[5] While performing the fitting routine, $I(Q,\omega)$ defined above must be inserted in Eq. (6.22) to build up the model lineshape $I_M(Q,\omega)$. The latter

---

[5]This method of accounting for the detailed balance does not apply when dealing with the extremely high (Q,ω)'s IA spectrum in Eq. (6.5). In fact, it can be easily verified that such a profile fulfills the detailed balance principle with no need of additional pre-factors.

contains some free parameters to be determined as a result of a best-fit procedure of the measured lineshape, based on the minimization of the following χ^2-function:

$$\chi^2 = \sum_{i=1}^{N} \frac{(I_M(Q,\omega_i) - y_i)^2}{\sigma_i^2}, \qquad (6.24)$$

where N is the number of energy points in the *IXS* spectrum, y_i is the number of counts at the energy $\hbar\omega_i$, σ_i is the related error and $I_M(Q,\omega_i)$ is the corresponding value of the model.

An appropriate metric to assess the overall quality of the fit is provided by the normalized distance between the model and the experimental measurement, namely, the normalized χ^2:

$$\chi_n^2 = \frac{\chi^2}{N - n_p} = \frac{1}{N - n_p} \sum_{i=1}^{N} \frac{(I_M(Q,\omega_i) - y_i)^2}{\sigma_i^2}. \qquad (6.25)$$

Here n_p represents the number of degrees of freedom of the best-fitting, i.e. the number of the parameters to be optimized, which can be written as $n_p = n_T - n_f$, with n_T and n_f being respectively the total number of the parameters in the model and the ones kept fixed to a value established as an external input.

In general, the determination of χ_n^2 requires an estimate of the count errors σ_i which might represent poorly the actual count fluctuations, thus ultimately resulting in a deceptive value of the minimum χ_n^2. This might owe, for instance, to some neglected source of error, relating to the nature of count detection. The use of χ_n^2 as a reference parameter is particularly useful as its expected value, $E(\chi_n^2) = 1$, is known and rather straightforward. However, a minimized χ_n^2 close to one is, by itself, not a reliable indicator of fitting accuracy.

In fact, if relative errors σ_i/y_i are significant, even a poor modelling, i.e. a systematic discrepancy between $I_M(Q,\omega_i)$ and y_i, causes small variations of the χ_n^2-factor $[I_M(Q,\omega_i) - y_i]^2/\sigma_i^2$. In other terms, best fitting routines performed on spectral measurements with poor statistical accuracy are more forgiving and tend to deliver small values of the optimized χ_n^2, regardless of possible model deficiencies.

From the above arguments, it follows that a small minimum χ_n^2 might provide a reliable indication of the modelling soundness only when the measurement has a high level of statistical accuracy.

In principle, to improve the agreement between the model and experimental data, one could increase the number of "degrees of freedom" (free parameters) contained in the fitting model. However, when such a number becomes excessive, best-fit values of the model parameters acquire high statistical correlation, thus becoming scarcely reliable.

A more rigorous evaluation of the "likelihood" of a specific model based upon a given experimental outcome could, in principle, be provided by a Bayesian inferential analysis, whose detailed discussion goes beyond the scope of this book. Bayesian methods (D'Agostini, 2003) have been sometimes used to interpret the shape of both quasielastic (Sivia and Carlile, 1992) and inelastic (De Francesco *et al.*, 2016) neutron spectrum of liquids. Briefly, these methods use extensive Monte Carlo simulations (Wang *et al.*, 2008) to establish whether the measured spectral shapes authorizes a particular hypothesis on the model. It can give a quantitative answer to questions concerning, for instance, the number of modes likely to contribute to the real spectrum or the likelihood of a given analytical form for the spectrum, based on the achieved. Of course, the ultimate scope of these Bayesian analyses is not to surrogating the role of the investigator, but rather providing an evidence-based probabilistic support for the choice of a given lineshape model.

Appendix 6A: The high and low-frequency limit of the memory function: The Damped Harmonic Oscillator model

We saw that, if internal rearrangements of the fluid occur over time lapses much shorter than any timescale relevant to the measurement, the time-decay of the viscous part of the memory function can be approximated by a Markovian term, that is a term proportional to $\delta(t)$. An example is provided by the viscous contribution to the hydrodynamic memory function in Eq. (6.4), which, indeed, has a

Markovian form reflecting the instantaneous character that liquid rearrangements assume when probed at hydrodynamic, or quasi-macroscopic, scales.

Conversely, for most liquids, thermal fluctuations are considerably slow and can be adequately described by a constant term in the time-dependent memory function. Consequently, the hydrodynamic memory function safely approximated by the $D_T \to 0$ limit of Eq. (6.4). Namely:

$$m_L(Q \to 0, t) = \omega_0^2(\gamma - 1) + 2\nu_L Q^2 \delta(t) \qquad (6A.1)$$

This expression of the memory function is ideally suited to describe the behaviour of a simple fluid, whose response is essentially "viscous" and thermal exchanges are infinitely slow.

We stress here that the use of the memory function defined above eventually leads to retrieving the following shape:

$$\left. \frac{S(Q,\omega)}{S(Q)} \right|_{Q,\omega \to 0} = \frac{1}{\pi} \left\{ \frac{\omega_0^2 (Q \to 0)\, \nu_L Q^2}{[\omega^2 - (c_s Q)^2]^2 + \omega^2 (\nu_L Q^2)^2} + \frac{\gamma - 1}{\gamma} \pi \delta(Q) \right\},$$

$$(6A.2)$$

where $\omega_0^2(Q \to 0) = c_s^2 Q^2/\gamma = c_T^2 Q^2$. Let us now consider the opposite case in which the response of the fluid is probed over timescales much shorter than internal rearrangements, *i.e.* in the elastic limit. Such nearly "frozen" rearrangements can be accounted for by including a constant term in the memory function. Within the same simplified assumption $D_T \to 0$, the memory function in this limit can be written as:

$$m_L(Q, t) = \omega_0^2(\gamma - 1) + \nu_{L,\infty} Q^2 + 2\Gamma_0(Q)\delta(t), \qquad (6A.3)$$

where $\nu_{L,\infty}$ is the elastic limit of the kinematic longitudinal viscosity. In the elastic regime, the viscous part of the memory assumes the form of a flat contribution, reflecting the nearly "frozen" nature of internal rearrangements. Furthermore, the term $\propto \delta(t)$ arises from the coupling of density fluctuations with the rapid microscopic dynamics associated, *e.g.*, to collisional events or, whenever relevant, to fast intramolecular degrees of freedom. The memory function introduced above, in principle holds validity for times much shorter than those of viscous "structural" processes yet much longer than the time lapses characteristic of the fast microscopic dynamics. For *IXS* measurements, fast rearrangements typically occur at 10^{-13} s, or shorter, timescales.

It appears that the two memory functions introduced above are isomorph, it is thus not surprising that the corresponding $S(Q, \omega)$ are also isomorph. In particular, upon neglecting the contribution of the Rayleigh mode, in the viscous and elastic regimes, the spectral shape assumes the same general form:

$$S(Q, \omega) = DHO(\omega) + B\delta(\omega), \qquad (6A.4)$$

where the first term is the so-called Damped Harmonic Oscillator (*DHO*) profile:

$$DHO(\omega) = \frac{2\Gamma(Q)\Omega^2(Q)}{[\omega^2 - \Omega^2(Q)]^2 + \Gamma^2(Q)\omega^2}. \qquad (6A.5)$$

This term accounts for the presence of two inelastic, Stokes and anti-Stokes, modes centred at the frequencies $\omega_s = \pm\sqrt{\Omega^2(Q) - [\Gamma(Q)/2]^2}$ and having damping coefficient $\Gamma(Q)$. A more in-depth discussion of the *DHO* model and its isomorphism with other 'popular" lineshape models is proposed in in Ref. (Bafile *et al.*, 2006).

We recall here that in Chapter 2 an explicit derivation of the *DHO* spectral shape was achieved in the framework of the linear response theory, by applying the main result of the fluctuation-dissipation theorem to the case of a classical (non-quantum) damped harmonic oscillator.

References

Bafile, U., Guarini, E. & Barocchi, F. 2006. *Phys. Rev. E* **73**, 061203.

Barocchi, F., Bafile, U., & Sampoli, M. 2012. *Phys. Rev. E* **85**, 022102.

Berne, B. J. & Pecora, R. 1976. *Dynamic Light Scattering.*, New York, Wiley.

Boon, J. P. & Yip, S. 1980. *Molecular Hydrodynamics*, Dover, Courier Dover Publications.

D'Agostini, G. 2003. *Rep. Prog. Phys.*, **66**, 1383.

De Francesco, A., Guarini, E., Bafile, U., Formisano, F. & Scaccia, L. 2016. *Phys. Rev. E* **94**, 023305.

Egelstaff, P. A. 1967. *An Introduction to the Liquid State*, London, New York, Academic Press.

Herzfeld, K. F. & Litovitz, T. A. 1965. *Absorption and Dispersion of Ultrasonic Waves*, London, Academic Press.

Landau, L. & Placzek, G. 1934. *Phys. Z. Sowjetunion*, **5**, 172.

Lewis, J. & Lovesey, S. 1977. *J. Phys. C* **10**, 3221.

Lovesey, S. 1971. *J. Phys. C* **4**, 3057.

Lovesey, S. W. 1984. *Theory of Neutron Scattering from Condensed Matter.* Oxford University Press, Oxford Vol. 1.

Sivia, D. & Carlile, C. 1992. *J. Chem. Phys.* **96**, 170.

Wang, Y., Zhou, X., Wang, H., Li, K., Yao, L. & Wong, S. T. 2008. *Bioinformatics,* **24**, 407.

Chapter 7

The Q-evolution of the spectral shape from the hydrodynamic to the kinetic regime

7.1. Using *THz* spectroscopy to detect mesoscopic collective modes: Early results

Broadly speaking, the dynamic response investigated by a spectroscopic measurement depends on the range of distances probed, or, in the reciprocal space, on the exchanged wavevector Q. Let us consider, for argument sake, a measurement aiming at characterizing the persistence of generalized hydrodynamic modes in the spectrum at finite Q values.

It is reasonable to expect that, while increasing Q, the lifetime of extended acoustic modes gradually shortens, due to decreasing ability of the system to support high-frequency collective propagation, and also diffusive motions become increasingly hindered by microscopic "in cage" interactions.

Furthermore, at high Q's, density fluctuations involve smaller volumes and fewer atoms; at some point, it can be expected that such fluctuations can no longer be described by a simple hydrodynamic theory, which applies to macroscopic or quasi-macroscopic scales only. Eventually, when the exchanged wavelength $\lambda_Q = 2\pi/Q$ decreases down to match — or become shorter than — first neighbouring atomic separations, the very notion of "collective mode" should, in principle, lose physical significance.

However, the case of fluids at liquid densities is more complicated than represented by this scenario. In fact, these systems are so tightly packed that the mean free path of their atoms (molecules),

169

l, becomes significantly smaller than the atomic size. Under these conditions, the short-time movements experienced by single atoms mostly reduce to "in cage" rattling oscillations having a period, T_r, in the sub-picosecond window. As a consequence, the conditions $2\pi/Q \gg l$ and $2\pi/\omega \gg T_r$ can be simultaneously fulfilled even at mesoscopic (Q, ω)'s. This means that, even in this dynamic regime, the response of a system is averaged over many elementary events and over distances largely exceeding those spanned by first neighbour atomic interactions, which is a pre-requisite for a suitably generalized hydrodynamic description of the dynamics.

Therefore, in the early 1960s, when *INS* technique was still in its infancy, the persistence of suitably generalized hydrodynamic modes down to mesoscopic scales appeared as a somehow intriguing, yet still plausible, hypothesis. This topic stimulated intensive experimental scrutiny, including a seminal measurement on dense Ne, Ar, and D_2 reporting the first evidence for an "extended hydrodynamic" shape of the spectrum (Chen *et al.*, 1965). The authors of this work reported that, at the lowest Q's, the frequency of inelastic peaks approached from above the known linear hydrodynamic trend, thus suggesting the hydrodynamic-like nature of these spectral features. However, this conclusion seemed partially at odds with the results of the *INS* work of Kroô and collaborators on Ar (Kroô *et al.*, 1964), which instead reported no evidence for distinct inelastic features. Nonetheless, even in this work, the presence of phonon-like excitations emerged indirectly from a comparison between liquid and solid phase spectra. Likewise, inconsistencies in these two early *INS* measurements owed to the different incident wavelength employed, which made the results hardly comparable.

In the same period, an *INS* measurement on Ar in both liquid ($T = 94$ K and 102 K) and solid ($T = 68$ K and 78 K) phases (Skold and Larsson, 1967) reported a linear Q-dependence of the inelastic shift at the lowest Q's, whose slope was consistent with the adiabatic sound velocity. Furthermore, at higher Q's the sound dispersion curve showed a vague resemblance with its counterpart in a crystal.

The presence of clear $S(Q, \omega)$ peaks at Q values as large as $16 \, \text{nm}^{-1}$ was also reported by an *INS* work on molten lead (Randolph and Singwi, 1966). A later *INS* measurement on the same sample (Dorner *et al.*, 1967) confirmed the existence of a vague similarity between the sound dispersion in fluids and the acoustic phonon branches of crystalline solids. Most importantly, this measurement also evidenced a typical solid-like behavior as the emergence of a transverse acoustic mode in the *INS* spectrum.

It is worth mentioning that previous *INS* measurements on superfluid He (Henshaw and Woods, 1961) had already evidenced solid-like features in the terahertz response, as indicated by the presence of well-resolved inelastic peaks exhibiting an approximately sinusoidal Q-dispersion. However, the quantum character of this system made the interpretation of results less straightforward and not obviously generalized to classical systems.

7.2. Evidence of extended Brillouin peaks at mesoscopic scales

Among earliest demonstrations of the persistence of a hydrodynamic-like shape well beyond the hydrodynamic regime, those achieved by two *INS* works on supercritical neon (Bell *et al.*, 1975, Bell *et al.*, 1973) and liquid Rb (Copley and Rowe, 1974) were especially compelling. Selected spectra measured in these works are respectively reported in Figure 7-1 and Figure 7-2, both evidencing the persistence of an unmistakable hydrodynamic triplet shape. In particular, the *INS* measurement by Bell and collaborators spanned Q's unusually low for neutron scattering measurements ($0.6\,\mathrm{nm}^{-1} \leq Q \leq 1.4\,\mathrm{nm}^{-1}$) in the attempt of reducing the dynamic gap with a hypersonic spectroscopic technique such as Brillouin Light Scattering (BLS). In Figure 7-1 these spectra are compared with those measured successively on supercritical Ar (Bafile *et al.*, 1990) in a Q range ($0.35\,\mathrm{nm}^{-1} \leq Q \leq 1.25\,\mathrm{nm}^{-1}$) further approaching the *BLS* Q-domain. Despite the effort of narrowing down the mentioned gap, the minimum Q value probed by these *INS* measurements still exceeded typical *BLS* Q range by more than one order of magnitude; in this respect, the persistence of an hydrodynamic-like triplet spectral shape was noteworthy.

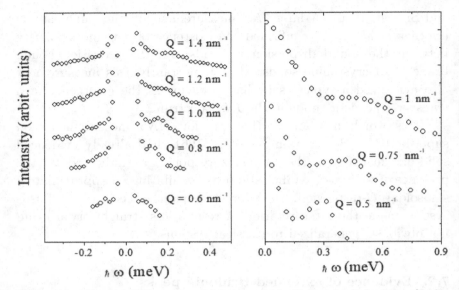

Figure 7-1:   Typical *INS* spectra of Ar measured by Bell and collaborators (Bafile *et al.*, 1990) (left plot) and by Bafile and collaborators (Bafile *et al.*, 1990) (right plot) at the indicated Q values. The intensities are vertically shifted for clarity.

The *INS* work on liquid Rb (Figure 7-2) covered Q values extending up to $10\,\mathrm{nm}^{-1}$, thus exceeding by nearly one order of magnitude the range spanned by Bell and collaborators.

At the time of these measurements, the persistence of a hydrodynamic-like spectral shape up to such high Q's appeared surprising, and the desire of shedding some light onto this observation motivated Bell and collaborators to perform a best-fit analysis of *INS* spectra with the generalized hydrodynamic model in Eq. (6.6). The values of the transport parameters extracted from such a lineshape analysis are reported in Figure 7-3 after normalization to the corresponding hydrodynamic predictions, as derived from Eq. (6.2), while using thermo-physical (NIST) and bulk viscosity data (Castillo and Castañeda, 1988), and the information included in the original references.

From Figure 7-3, it can be readily appreciated that these normalized transport parameters are very close to one, except perhaps the lowest Q data of Bafile and collaborators. This trend pinpoints

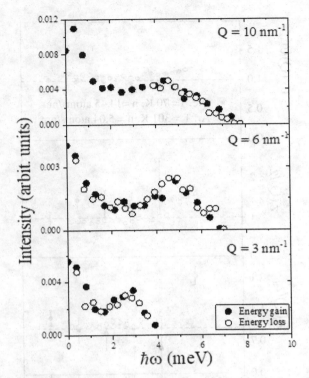

Figure 7-2: *INS* spectra of liquid Rb. All spectra are redrawn from Ref. (Copley and Rowe, 1974).

an overall consistency between hydrodynamic and mesoscopic values of transport parameters, which seems even more unexpected as one considers the highly damped character of spectral modes in Figure 7-1. The relatively large damping of the extended hydrodynamic modes indicates a substantial decrease in their lifetime. This evidence, at odds with data in Fig. 7-3, would rather suggest a failure in the simple hydrodynamic predictions of Eq. (6.2), which rest on the assumption of long-living spectral modes ($D_T Q^2$, $\Gamma Q^2 \ll c_s Q$).

For these reasons, these low-Q *INS* measurements on noble gases were considered breakthrough yet somehow baffling results, demonstrating the persistence of "extended hydrodynamic" modes down to mesoscopic scales (Bafile *et al.*, 1990).

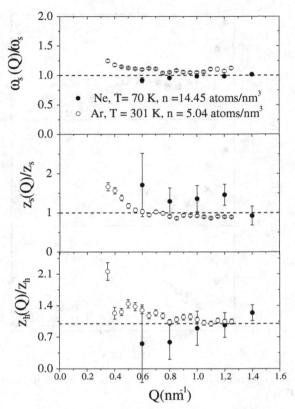

Figure 7-3:   Shape parameters of the generalized hydrodynamic model of $S(Q, \omega)$ (see Eq. (6.6)) of Ne (dots) and Ar (circles) at the indicated thermodynamic conditions, as reported in Refs. (Bell *et al.*, 1975) and (Bafile *et al.*, 1990), respectively. Data are normalized to the corresponding hydrodynamic values as derived by inserting in Eq. (6.2) transport parameters either extracted from the original works, or from a database of the National Institute of Standards and Technology (NIST) and bulk viscosity data in Ref. (Castillo and Castañeda, 1988).

Up to what Q value does a hydrodynamic-like behavior persist?

At this stage, a question arises on the existence of Q-threshold above which the breakdown of the simple hydrodynamic behavior occurs. As discussed in Chapter 8 and Chapter 9, a substantial amount of successive *IXS* results proves that the persistence or not of a hydrodynamic behavior depends non-trivially on the thermodynamic conditions of the sample. In this respect, it is worth noticing that the spectra in Figure 7-3 refer to deeply supercritical or gaseous samples, for which viscoelastic effects leading to non-hydrodynamic deviations at low-to-moderate Q's are relatively weak.

As discussed more extensively in Chapter 8, the behavior of a subcritical (liquid) sample is entirely different, as its generalized (Q-dependent) sound velocity sharply increases with Q, while the corresponding longitudinal viscosity decreases. These trends mirror the transition from a "liquid-like" behavior and to "solid-like" one. An example of this trend is provided in Figure 7-4, which illustrates the influence of these effects on the generalized sound velocity, as derived from *INS* measurements on liquid Ar at various pressures (de Schepper *et al.*, 1984b). The plots show that, up to intermediate Q's, the sound velocity derived from a terahertz spectroscopic measurement largely exceeds its hydrodynamic value, c_s. Furthermore, all plots show that, after reaching some Q-maximum or plateau between $5 \, \mathrm{nm}^{-1}$ and $8 \, \mathrm{nm}^{-1}$, $c_s(Q)$ starts to decrease; this trend is due to the onset of destructive interference between the acoustic wave and the local pseudo-periodical arrangement of the atoms.

Structure and dynamics

Overall, the *INS* measurements discussed thus far demonstrated that when the probed wavelength, $2\pi/Q$, approaches the average distance between adjacent atoms, the propagation of collective excitations become strongly coupled with the local structure of the fluid. The variable ideally suited to characterize such a coupling is the static structure factor (see Chapter 2, Section 2.8) which, for simple monatomic liquids, assumes the typical shape reported in Figure 7-5, as measured by X-Ray diffraction in liquid caesium (Ohse, 1985).

Qualitatively, the $S(Q)$ of a simple liquid has a profile featuring a sharp maximum at $Q = Q_m \approx 2\pi/a$ where a is a characteristic distance of the fluid somehow close to the average adjacent atoms'

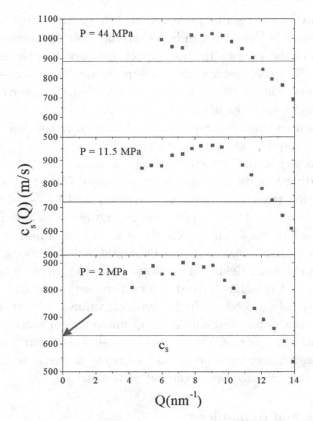

Figure 7-4:   The Q-dependence of the generalized sound velocity, $c_s(Q)$, derived from best-fit values of the shift parameter, $\omega_s(Q)$, in the lineshape model of Eq. (6.6). Data extracted from *INS* measurements of Ref. (de Schepper *et al.*, 1984b) on liquid Ar at the indicated pressures. The horizontal lines show the corresponding low Q hydrodynamic limit, i.e. the adiabatic sound velocity of the sample. Finally, the arrow in the bottom panel indicates the $Q = 0$ limit of the curve.

separation. Notice that the actual separation distance between first neighbouring atoms can be more rigorously determined as the position of the first maximum in the pair distribution function introduced in Eq. (2.81) (see, e.g., Egelstaff, 1967). As discussed further below in this chapter, beyond the first diffraction peak, $S(Q)$ displays Q-oscillations which gradually damp off upon increasing

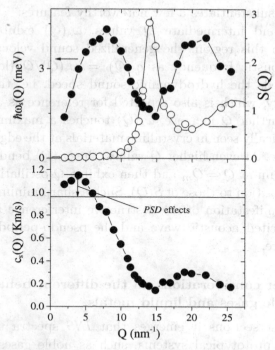

Figure 7-5: In the lower plot, the generalized sound velocity $c_s(Q)$ of liquid cesium is reported (line + open dots) as derived from *INS* measurements in Ref. (Bodensteiner *et al.*, 1992), where the horizontal line represents the hydrodynamic value. The amplitude of the positive sound dispersion (*PSD*) of $c_s(Q)$ is also schematically shown for reference. The upper plot compares the corresponding generalized sound dispersion $\omega_s(Q)$ (line + dots) along with the static structure factors $S(Q)$ therein derived from the INS spectra in the same work (line + open circles) or from X-Ray diffraction measurements (solid line, (Ohse, 1985)).

Q, until the $S(Q) = 1$ incoherent limit is reached at extremely high Q values. In this regime, the signal probed by a spectroscopic measurement mainly couples with the free recoil of the atom struck in the scattering event, as discussed in the previous chapter.

In the upper plot of Figure 7-5, the $S(Q)$ of caesium is compared with the dispersion curve, i.e. the Q dependence of the inelastic shift, $\omega_s(Q)$, extracted from the *INS* spectrum of the same material (Bodensteiner *et al.*, 1992).

The plot substantiates a few noteworthy features:

At low and intermediate Q values, $\omega_s(Q)$ exhibits a steep Q-growth. In this region, the generalized sound velocity deduced from the acoustic frequency as $c_s(Q) = \omega_s(Q)/Q$ (lower graph) slightly exceeds the hydrodynamic sound speed, i.e. the adiabatic sound velocity, which is also reported for reference as a horizontal line. Upon further Q increase, $\omega_s(Q)$ touches a maximum at $Q = Q_m/2$, as typically seen in crystalline materials at the edge of the first Brillouin zone. At even higher Q values, the $\omega_s(Q)$ bends down to a sharp minimum at $Q = Q_m$ and then exhibit Q-oscillations roughly in phase opposition to those of $S(Q)$. Such a sharp minimum appears as a clear manifestation of the destructive interference between the scattering-excited acoustic wave and the pseudo-periodicity of the local structure.

7.3. Further considerations on the different behavior of noble gases and liquid metals

From the last sections it emerges that *INS* spectra measured in the simplest prototypical systems, such as noble gases (neon and argon) differ in many aspects from those measured in molten metals (rubidium and caesium). The latter systems are in principle more complex, due to the influence of Coulomb forces on atomic interactions. Some relevant differences are discussed below.

The persistence of a hydrodynamic-like shape

While inelastic modes in neon spectra already become highly damped at $Q = 1\,\mathrm{nm}^{-1}$, for liquid rubidium they are still clearly discernible up to a Q value as high as $10\,\mathrm{nm}^{-1}$. This discrepancy has been sometimes ascribed to the different amplitude of the repulsive part of the interatomic potential. In noble gases, which represent more closely the ideal case of hard-sphere systems, such a repulsive portion is much more pronounced than in liquid metals (Balucani *et al.*, 1993). To achieve a more quantitative assessment of these differences, it is useful to refer to the distance between first neighbouring atoms, *d*. Indeed, one can estimate that in hard-sphere models inelastic peaks survive until $Qd = 0.5$, in liquid Ar until $Qd \approx 1$ and in liquid Rb until $Qd \approx 4$ (de Schepper and Cohen, 1980).

The positive sound dispersion

One of the most striking dispersive effects discussed in the previous sections is the presence of a systematic Q-increase of sound velocity at the departure from the hydrodynamic limit. Such effect, customarily referred to as the positive sound dispersion (PSD), will be discussed in further detail in Chapter 8, which will also clarify its link with the viscoelasticity of the fluid. A comparison between Figure 7-4 and the lowest plot of Figure 7-5 is consistent with what emerges from the literature, i.e. that the amplitude of PSD effects is larger in liquid noble gases than in molten metals.

A large Einstein frequency is often considered as a rough predictor of the persistence of inelastic modes up to large Q's and the occurrence of a large PSD. As readily inferred from Eq. (2.21), it appears that Ω_0 directly relates to the average Laplacian of potential, whose dominant contribution comes from the repulsive region, where the potential has the steepest variation with r. Consistently, INS (de Schepper *et al.* 1984a,b, van Well and de Graaf 1985, van Well *et al.* 1985) clearly demonstrated that the PSD is more pronounced in noble gases systems, for which this repulsive part is especially harsh.

The point of view of *MD* simulations

An ideal tool to investigate the response of simple fluids are Molecular Dynamics (MD) computer simulations. Indeed, interparticle interactions in simple systems can be mimicked more easily by model potentials. As an example, an MD study on a model of soft spheres (Kambayashi and Hiwatari, 1994) confirmed that the spectra from systems having softer interaction potentials bear evidence of distinct acoustic-like modes up to larger Q values. This result might explain why distinct side peaks in the $S(Q,\omega)$ of liquid metals persist up to Q values substantially larger than for noble gases.

In *MD* simulations, the different microscopic interactions of rare gases and molten metals justify the use of different models for the interatomic potential. For noble gases, the model usually preferred is a Lennard–Jones (*LJ*) potential:

$$\Phi(r) = 4\varepsilon[(\sigma/r)^{12} - (\sigma/r)^{6}].$$

In this model, ε represents the depth of the potential well, σ the distance corresponding to the first zero of the *LJ* potential, r the distance between two interacting atoms, and $r_m \approx 1.122\sigma$ is the distance at which the potential reaches its minimum value, namely $-\varepsilon$.

The actual values of these parameters can be tailored to reproduce thermodynamic or transport properties of the system to be simulated. The repulsive term, i.e. the one $\propto r^{-12}$, accounts for the short-range Pauli repulsion arising from the overlap between electronic orbitals; the term $\propto r^{-6}$ accounts instead for long-range van der Waals attractive forces.

The model for the interatomic potential of molten metals is slightly more complex, due to the presence of the conduction electrons. However, pseudopotential theories (Cohen, 1986) describe the behavior of ions in liquid metals by considering a simple one-component fluid of electrically neutral pseudo-atoms, which interact through an effective liquid metal (*LM*) potential. The repulsive cores of these effective interactions are softer than the *LJ* ones. Furthermore, at variance with the *LJ* potential, the *LM* one has an oscillatory tail due to the conduction electrons.

Although differences in the structural properties of *LJ* and *LM* models are often marginal, those observed in the dynamic properties are substantial and still not completely understood.

The solid-like high-frequency response of liquid metals

In summary, the pronounced "phonon-like" character of acoustic modes in molten metals stems from the circumstance that soft-core interactions increase the coherence of collective atomic motions. For instance, it is commonly observed that the velocity autocorrelation functions in *LM* computer models show more pronounced oscillations than their *LJ* counterparts. Also, it seems well-assessed that *LM* models more efficiently support shear wave propagation. Along this line, it is worth mentioning a computational work of Mountain (Mountain, 1982) on a model of liquid rubidium. In this work, the power spectra of the transverse-current correlation function were used to determine up to what distance the hydrodynamic prediction of an exponential time-decay holds validity. Mountain found that this "hydrodynamic length", measuring the spatial extent of transverse dynamical correlations, increases significantly with the amount of supercooling.

A model for liquid rubidium was also considered in a successive computer simulation by Balucani and collaborators (Balucani *et al.*, 1987), dealing with the wavevector dependence of the transverse current memory. In essence, this variable is simply related to the generalized shear viscosity, $\eta_s(Q,t)$ as $m_L(Q,t)$ is related to $\eta_L(Q,t)$. The authors reported that, at the crossover between hydrodynamic and kinetic regimes, the main inelastic feature dominating this generalized shear viscosity is much more pronounced in Rb than in Lennard-Jones systems, consistently with the discussed differences in the transport properties of liquid metals and simpler systems, such as noble gases.

The simulation work described in Ref. (Canales and Padró, 1999) proposes a comparative study of Lennard Jones and a liquid metal model, in which both single-particle and collective dynamic response functions were considered, including transverse current spectra. Figure 7-6 highlights the different abilities of these two model systems in supporting shear mode propagation. For a liquid metal, a discernible inelastic peak in the current spectra already emerges at $Q = 2.5\,\text{nm}^{-1}$, while for a Lennard–Jones model at the same Q the same spectral function has the form of a monotonically decaying peak centred at $\omega = 0$. Furthermore, at the highest Q value, it can be noticed that the peak dominating the transverse current of the *LM* model is much more pronounced than for the *LJ* one.

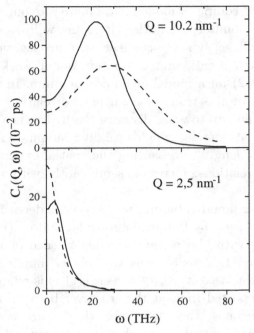

Figure 7-6:   The transverse current spectra of a computer model representative of a molten metal (solid lines) is compared to the one obtained for a Lennard-Jones (dashed lines), for the Q values indicated in the plots.

7.4. The kinetic theory approach: A few introductory topics

Among the aims pursued by various *INS* works carried out before the advent of the first *IXS* spectrometers, the experimental validation of the Kinetic Theory (*KT*) predictions has played a central role.

The *KT* is a statistical mechanics approach initially developed to provide a rigorous account of the microscopic dynamics of fluids within the hypothesis of uncorrelated collisions (Cohen, 1993, Yip, 1979). Considering the close-contact nature of these interactions, it is not surprising that long-range correlations typical of liquid phase aggregates are not adequately accounted for by this theory. This limitation has practically restricted the domain of validity of *KT* to hard-sphere-like systems, i.e. systems in which the short-range repulsive part of the potential is dominating; consequently, kinetic theories, in principle, rigorously apply to dilute gases only. Nonetheless, KTs provide a rigorous and non-phenomenological account of the dynamical process occurring inside a many-body system. In this framework, the response is represented in terms of phase-space variables, i.e. variables dependent on both atomic positions and velocities, as required by a consistent treatment of microscopic collisions.

A generalization of the original Boltzmann–Enskog description customarily referred to as the revised Enskog theory, made it better suited for dense fluids by assuming a density-dependence in the Enskog operator. Despite some advances prompted in the mid-1970s by this theoretical improvement, KT approaches have been for long trapped in an impasse, mainly due to the impossibility of expressing transport parameters as a power series of the density (Cohen, 1993), as usually done for thermodynamic coefficients. However, the successive advent of extensive computational investigations substantially revitalized this approach.

Since the KT treatment of correlation functions uses a larger number of atomic coordinates, it is formally more complicated than ordinary time-space descriptions. Nevertheless, it provides a more rigorous theoretical framework to derive quantities of relevance for spectroscopy experiments, such as static and dynamic structure factors. A more formal introduction of a few relevant aspects of the kinetic theory can be found in Appendix 7A; we only mention here that such a phase space description enables the explicit derivation of single-particle correlation spectrum $S(Q, \mathbf{v}, \mathbf{v}', \omega)$, which not only depends on the dynamic variables (Q, ω) but also on the initial ($\mathbf{v}$) and final ($\mathbf{v}'$) velocities of the colliding particles. The standard dynamic structure factor $S(Q, \omega)$ is finally obtained from the double integration of the phase space spectrum $S(Q, \mathbf{v}, \mathbf{v}', \omega)$ over the two variables (v) and $\mathbf{v}'$ of all atoms in the system.

Among various experimental attempts to validate the KT predictions on the spectral lineshape, it is worth mentioning an INS work on Ar by de Schepper and collaborators (de Schepper *et al.*, 1984b) which aimed at testing the expansion of $S(Q, \boldsymbol{\omega})$ as a sum of generalized Lorentzian terms (see Appendix 7A). In this work, the effect of spectral rule fulfilment was also considered by comparing the lineshape modeling obtained by imposing sum rules of increasing order (Ernst *et al.*, 1969). More generally, similar spectral shape models have been successfully tested in various experimental works on noble gases (van Well *et al.*, 1985, van Well and de Graaf, 1985, de Schepper *et al.*, 1984a, Verkerk and van Well, 1986).

Toward the end of the 1970s, Postol and Pelizzari (Postol and Pelizzari, 1978) used *INS* spectra to benchmark *KT* predictions for the dynamic structure factor. Figure 7-7 propose a comparison between a constant Q lineshape measured by Postol and Pelizzari and the corresponding *KT* profile. It appears that the agreement between theoretical predictions and experimental results was, at best, merely qualitative.

The experimental validation of *KT* predictions was also the focus of a joint *INS* and computational work on krypton at various densities along a room temperature isothermal path (Egelstaff *et al.*,

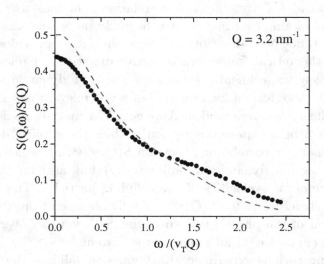

Figure 7-7:   *INS* spectrum of argon gas (dots) compared with the corresponding kinetic theory prediction (dashed line). Redrawn from (Postol and Pelizzari, 1978).

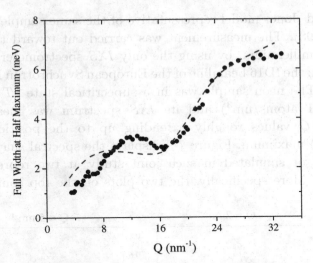

Figure 7-8: FWHM of *INS* spectra of krypton measured by Egelstaff and collaborators (Egelstaff *et al.*, 1983).

1983). The neutron time-of-flight spectra were measured at the IN4 spectrometer at the high-flux reactor of the Institut Laue–Langevin. The *FWHM* of the essentially featureless *INS* spectra measured by Egelstaff and collaborators is compared with its *KT* prediction in Figure 7-8. Although the agreement between the experimental and theoretical curves seems satisfactory at high *Q*'s, evident discrepancies are observed at moderate to low exchanged wavevectors where collective modes dominate density fluctuations.

7.5. The onset of kinetic regime probed by *IXS* measurements on deeply supercritical neon

Before the development of the first high-resolution *IXS* spectrometer in the mid-1990s, the transition from the (extended) hydrodynamic regime to the kinetic one was considered a topic of prominent scientific interest, certainly worth intensive investigative efforts. Along the pathway traced by past *INS* experiments, an early *IXS* measurement on deeply supercritical neon aimed at a precise characterization of this crossover by using the support of a parallel *MD* simulation on a Lennard–Jones model representative of the same sample (Cunsolo *et al.*, 1998). The measurement was carried out toward the end of the past millennium, by using the only *IXS* spectrometer available back then: the ID16 beamline of the European Synchrotron Radiation Facility. The neon sample was in a supercritical state ($T = 295$ K, $n = 29.1$ atoms/nm³) and its *IXS* spectrum was measured at different Q values roughly extending up to the position of the first $S(Q)$ maximum. Figure 7-9 displays the spectral shapes either measured or simulated in such joint study at two representative Q values. More specifically, the two plots on the top compare raw

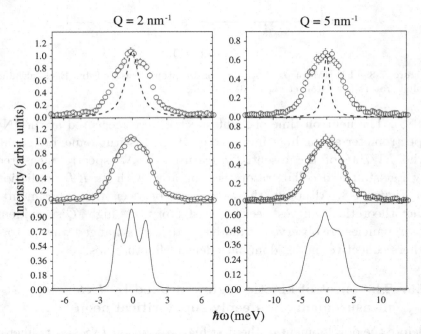

Figure 7-9: Two top plots: *IXS* spectra measured, at the indicated Q values, on deeply supercritical neon ($T = 295$ K, $n = 29.1$ atoms/nm³, open circles) are compared with corresponding resolution functions (dashed lines). Two bottom plots: MD spectra computed at the same Q values on a Lennard Jones model representative of the same neon sample (solid lines). Two middle plots: Comparison between *IXS* spectra (open circles) and the corresponding molecular dynamics (MD) ones convoluted with the resolution functions after multiplication by the detailed balance factor (see the end of Chapter 6) (solid lines). Data are redrawn from Ref. (Cunsolo *et al.*, 1998).

IXS spectra with the corresponding energy resolution profile, while the two bottom ones show the lineshapes simulated at the same Q's. Finally, the middle panels compare simulated spectra to the measured ones after the latter are corrected for the detailed balance factor (see Section 6.3) and convoluted with the energy resolution profile.

The substantial agreement between measured and computed lineshapes suggests that the Lennard–Jones model potential provides a consistent description of the *IXS* spectrum of deeply supercritical neon. More importantly, the simulated spectra show the persistence of an hydrodynamic-like triplet shape down to mesoscopic distances (see bottom plots), as previously evidenced, in a smaller Q range, by *INS* spectra of noble gases in Figure 7-1.

At this stage, it still remains an unclarified question concerning the Q-threshold below which the spectral shape of a simple system keeps its "extended hydrodynamic" shape: according to the prediction of Kamgar–Parsi and collaborators (Kamgar–Parsi *et al.*, 1987, see also Appendix 7A), this Q-threshold should be comparable with the inverse of the Enskog mean free path. The accuracy of this prediction can be judged from Figure 7-10, which displays the Q-dependence of the generalized transport parameters derived from the best-fit modeling of *MD* and *IXS* lineshapes reported in Ref. (Cunsolo, Pratesi *et al.*, 1998), carried out using the *GH* model in Eq. (6.6). Indeed, consistent with *INS* results in Figure 7-3, all transport parameters derived from this lineshape modeling join at lowest Q's the respective hydrodynamic predictions, although somehow less clearly so for the thermal diffusivity coefficient. This extended hydrodynamic behavior ceases at larger Q's where the onset of a Q-dependence of transport parameters becomes evident.

A rough boundary between this "extended hydrodynamic region" and the higher Q regime, here dubbed "kinetic", is roughly delimited by the shadowed vertical band centred to about $8\,nm^{-1}$, which is a value not too distant from the predicted crossover to the kinetic regime, with condition $Ql_E \gg 1$, where l_E — about $10\,nm^{-1}$ in the present case — is the Enskog mean free path (see Appendix 7A).

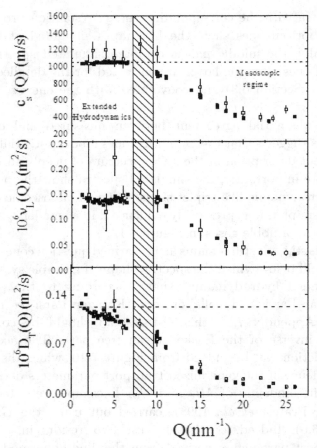

Figure 7-10:   The relevant transport parameters deduced from the fits of *IXS* spectra of deeply supercritical neon (open squares) and those of MD spectra from a Lennard Jones model representative of the same sample (full squares). Data are from Refs. (Cunsolo *et al.*, 1998) and (Cunsolo, 1999). The shadowed area roughly locates the transition from the extended hydrodynamics and the mesoscopic regime. Horizontal dashed lines represent the macroscopic (hydrodynamic) values deduced from thermodynamic (NIST) and transport properties (Castillo and Castañeda, 1988) tabulated in the literature.

Data reported in Figure 7-10 suggest the occurrence of a somehow smooth transition between the "extended hydrodynamic" and the "kinetic" regime with essentially no intermediate (viscoelastic) region in between them. This viscoelastic region, when occurring, is characterized by a Q increase of sound velocity, which is to be excluded based on data in the upper panel of Figure 7-10. As mentioned, this trend is far from being universal for all fluids, as one generally observes for liquid phase samples a more complex Q-dependence arising from the coupling with active relaxation processes (see Chapters 8 and 9).

7.6. The crossover from the collective to the single-particle regime: Some qualitative aspects

Historically, a recurring objective of spectroscopy investigations extending well beyond the hydrodynamic regime has been the study of the crossover to the single-particle limit. Figure 7-11 portrays a few relevant qualitative aspects associated to the spectral evolution throughout this large dynamic crossover by displaying the *IXS* spectra measured on lithium over a considerably broad Q-range (Scopigno *et al.*, 2000). For clarity, spectral shapes reported in the plot are vertically shifted by an amount roughly increasing with Q.

It can be noticed that, at the lowest Q's, the spectral shape consists of a relatively sharp triplet profile dominated by the three generalized hydrodynamic modes. The inelastic shift of the Brillouin side peaks increases somewhat linearly with Q, as pictorially suggested by the dashed line roughly connecting their maxima in the Stokes side.

Upon further Q-increase, the extended Brillouin peaks become more damped, gradually transforming into broad shoulders, almost submerged by the wings of the dominating central peak.

At even higher Q-values, the spectrum covers progressively broader energy-ranges, and, due to the detailed balance factor, becomes increasingly depleted in the anti-Stokes side. Due to the concomitant disappearance of the central peak, the spectral shape is increasingly skewed towards the Stokes side. The position of the only surviving feature on the Stokes side gradually approaches the limiting value $\omega_r = \hbar Q^2/2M$, while the overall spectral shape evolves

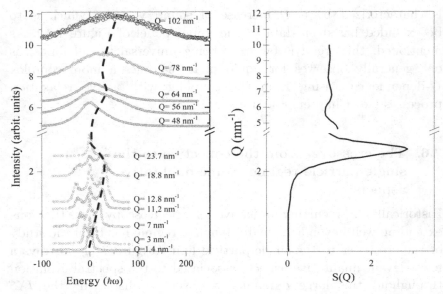

Figure 7-11:   The left plot illustrates the transition of the spectral shape from the collective to the single-particle regime, as measured in lithium at $T = 475\,\text{K}$. The thick dashed line serves as a guide to the eye roughly indicating the Q-evolution of the shift of the dominant inelastic feature on the Stokes side of the spectrum. In the right plot, the corresponding $S(Q)$ is reported for reference. The *IXS* spectra and the $S(Q)$ are redrawn from Ref. (Scopigno, Balucani *et al.*, 2000), and the former ones are vertically shifted by an amount roughly increasing as the corresponding Q value for clarity.

toward the Gaussian, single-particle profile. As a reference, the right plot of the same figure shows the parallel Q-dependence of the $S(Q)$ profile. As expected, the evolution of the lineshape toward the single-particle limit is reflected by the corresponding damping of the coherent Q-oscillations in the static structure factor.

Figure 7-12 provides a perhaps more complete portrait of the described spectral shape evolution at the crossover between hydrodynamic and single-particle regimes, as it emerges from the "patchwork" of few literature measurements on simple fluids. The static structure factor of lithium is also reported for reference in the central plot to identify the various Q-windows involved.

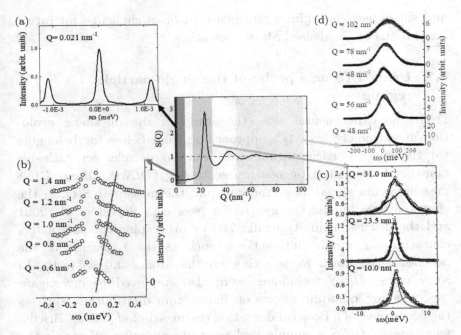

Figure 7-12: Overview of experimental spectra measured in several Q windows across the transition from the hydrodynamic to the single-particle regime in monatomic fluids. Panel (a) reports the Brillouin light scattering spectra of liquid Ar (Fleury and Boon 1969). Panel (b) displays *INS* measurements on supercritical Ne (Bell, Kollmar *et al.*, 1973) with the solid line roughly indicating the linear dispersion of side peaks. Panel (c) shows *IXS* spectra on liquid Ne (Cunsolo, Pratesi *et al.*, 2001) with corresponding best-fitting lineshapes and individual spectral components. Finally, Panel (d) displays *IXS* spectra of liquid Li from Ref. (Scopigno, Balucani *et al.*, 2000) along with the single-particle Gaussian profile as a solid line, almost completely hidden by overlapping measurements. The central panel reports the simulated $S(Q)$ of lithium from Ref. (Scopigno, Balucani *et al.*, 2000). Finally, the arrows and the shadowed areas in the central plot serve to identify the Q-range of each measurement.

One readily recognizes the various regimes of the lineshape already illustrated in Figure 7-11.

Besides this merely qualitative understanding, a quantitative and unified description of the spectral profile between hydrodynamic and single-particle regimes represents an open challenge for future generations of Condensed Matter scientists.

7.7. Using *IXS* as a probe of the single-particle regime

From the experimental side, the study of the lineshape evolution at extremely high Q's approaching from below (or belonging to) the single-particle regime has been for decades an exclusive domain of deep inelastic neutron scattering (*DINS*). Often, *DINS* measurements aimed at extracting from the measured $S(Q,\omega)$ the single-particle kinetic energy $\langle K.E.\rangle$ (see, e.g., Senesi *et al.*, 2001 and the discussion in Appendix 7B) or other physical parameters characterizing atomic interactions, such as the Laplacian of the potential, the mean force acting on the atom, and so forth. In some cases, *DINS* techniques were also employed to investigate "spectacular" quantum effects on momentum distribution, such as the occurrence of a Bose condensation (Stringari *et al.*, 2002). Briefly, what makes *DINS* a unique tool to study quantum effects, is the unique opportunity of accessing the extreme Q values where these effects become best observable (Silver and Sokol, 2013).

The use of *IXS* as a complementary probe of quantum effects was pioneered by a deep inelastic X-Ray scattering (*DIXS*) measurement on liquid neon at 26 K (Monaco *et al.*, 2002). Besides demonstrating the overall consistency between *DINS* and *DIXS* lineshapes, this work also evidenced the onset of substantial quantum effects on $\langle K.E.\rangle$. Measurement were performed at the 10ID beamline of the *ESRF*, while using the vertically rotating spectrometer arm to cover the unusually broad angular range required for this measurement; this corresponded to a maximum Q value as high as $160\,\mathrm{nm}^{-1}$, for the considered incident energy of 17,794 eV.

Figure 7-13: illustrates the Q-evolution of the spectrum toward the *IA* Gaussian shape, as it emerges from *DIXS* measurements of Monaco and collaborators. The plot shows that the ω-shift of *IXS* lineshape centroid systematically enhances upon Q-increase,

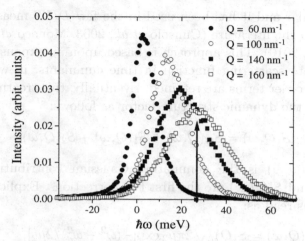

Figure 7-13: The Q-evolution of the *IXS* spectrum of Ne toward the Gaussian shape centred at the recoil energy characteristic of the single-particle regime. The *IXS* spectra, measured at the Q values shown in the plot, are taken from Ref. (Monaco, Cunsolo *et al.*, 2002). The vertical arrow shows the recoil energy computed for the highest Q spectrum.

asymptotically approaching the value of the recoil frequency ω_r also located by the vertical arrow for the highest Q spectrum.

7.8. Final states effects

As discussed in the previous section, in the single-particle or impulse approximation (*IA*) regime, the incident particle, e.g. an X-Ray photon, strikes the atom hard enough to free it from any interactions with neighbouring atoms and triggers its ballistic recoil.

For (Q, ω) values large, yet still smaller than those typical of the *IA* regime, the struck atom cannot be considered as a freely recoiling particle, as first neighbour interactions become relevant. It is, however, safe to assume that these interactions mainly influence the final state of the struck atom, its initial (before scattering) state still being free.

Among various analytical attempts to handle these "final state effects", the so-called *additive approach* (Glyde, 1994) is probably the most popular and it has been used in the few *DIXS* measurements reported in the literature (Cunsolo *et al.*, 2003, Monaco *et al.*, 2002, Izzo *et al.*, 2010). This approach is based upon an expansion of the intermediate scattering function in time cumulants, in which only the lowest order terms are retained. Eventually, this treatment leads to express the dynamic structure factor as follows:

$$S(Q, \omega) = S_G(Q, \omega) + S_1(Q, \omega) + S_2(Q, \omega), \tag{7.1}$$

where $S_G(Q, \omega)$ is the dominating Gaussian contribution, while $S_1(Q, \omega)$ and $S_2(Q, \omega)$ are the first two corrections. Explicitly, these three terms read as:

$$S_G(Q, \omega) = S(Q)/\sqrt{2\pi\mu_2} \exp[-(\omega^2 - \omega_d^2)/2\mu_2], \tag{7.2}$$

$$S_1(Q, \omega) = -\mu_3/8\mu_2^2(\omega - \omega_d)(1 - \omega_d^2/3\mu_2)S_G(Q, \omega), \tag{7.3}$$

$$S_2(Q, \omega) = -\mu_4/8\mu_2^2(1 - 2\omega_d^2/\mu_2 + \omega_d^4/3\mu_2^2)S_G(Q, \omega), \tag{7.4}$$

where the shift $\omega_d = \omega_r/S(Q)$ tends to the recoil frequency $\hbar Q^2 2M$ in the incoherent approximation $(S(Q) = 1)$ regime.

This perturbative treatment links deviations from a perfect Gaussian shape to the lowest order central spectral moments $\mu_n = \int_{-\infty}^{\infty} d\omega(\omega - \omega_r)^n S(Q, \omega)$; these spectral parameters depend on physically relevant variables such as the mean single particle kinetic energy, the mean force acting on the atom and the Laplacian of the potential, all of them providing a direct measure of the weight of quantum effects.

In the cited work of Monaco and collaborators, an additive approach based model like the one in Eqs. (7.1)–(7.4) was tested against the *IXS* lineshapes measured on liquid neon in the 10–160 nm^{-1} range. As a result, it was shown therein that only the first two lowest order corrections provide a sizable contribution to the spectral shape, their most visible signature being a slight asymmetry of the spectral tails with respect to the spectrum centroid. Also, this lineshape modeling led to derive a mean single atom kinetic energy exceeding its classical prediction by more than 20%; this result appeared consistent with previous *INS* determinations in the literature.

7.9. The case of molecular systems

After the first work by Monaco and collaborators on liquid neon, a similar *DIXS* measurement was performed on a diatomic homonuclear system — liquid iodine (Izzo *et al.*, 2010). Despite its slightly more complex microscopic constituents, this sample was in principle slightly simpler than liquid neon due to its entirely classical character, which owed to both the larger molecular mass and the high temperature ($T = 388$ K) probed in the measurement.

Even in the absence of sizable quantum effects, the interpretation of the *IXS* spectrum of a molecular fluid is often overly complicated, due to the coupling of the spectroscopic probe with all molecular degrees of freedom and their possible mutual entanglement (Krieger and Nelkin, 1957). In general, the spectral response of these systems depends on how $\hbar\omega$ compares, not only with molecular centres of mass translational energies, E_t, but also with intramolecular rotational and vibrational quanta, $\hbar\Omega_r$ and $\hbar\Omega_v$ respectively. Within the simplified assumption that various molecular degrees of freedom are mutually decoupled and have very different timescales, three complementary *IA* regimes can be identified:

(1) When $E_t \ll \hbar\omega \ll \hbar\Omega_r, \hbar\Omega_v$ (**translational *IA***) the struck molecule is "perceived" by the probe as a freely recoiling spherical particle. The energy of its translational recoil is equal to $\hbar^2 Q^2 / 2M$, $\hbar^2 Q^2 / 2M_t$ where M_t coincides with the mass of the whole molecule. Here, the scattered intensity carries direct insight into the purely translational momentum distribution of the molecular centres of mass.

(2) In the intermediate Sachs–Teller (*ST*) regime corresponding to $E_t, \hbar\Omega_r \ll E \ll \hbar\Omega_v$ (**roto-translational *IA***) the molecule is perceived by the probe as a recoiling rigid roto-translator. The rotational part of its recoil being $\hbar^2 Q^2 / 2M_{\mathrm{ST}}$, where M_{ST} is

the effective or ST mass, which is determined by the eigenvalues of the molecular tensors of inertia. In this regime, the spectral density becomes directly related to the roto-translational momentum distribution of the system.

(3) Eventually, when the $\hbar\omega \gg \hbar\Omega_v$ window (**vibrational IA**) is joined, the exchanged energy largely surpasses any inter- and intramolecular energy, and the scattering impact for short-time "frees" the nucleus of the molecule from its bound state, causing its free recoil. Under these conditions, the scattering intensity becomes proportional to the momentum distribution of the single freed atomic nucleus.

Finally, higher-level IA regimes can be probed at even larger exchanged energies (and momenta), which overpowers increasingly strong interactions. Eventually, when they far exceed intranuclear bonding energies, sub-nuclear particles are freed from their nuclear bonds and the study of their behavior belongs to the realm of Nuclear and Particle Physics.

Coming back to less extreme portions of the dynamic plane, the *DIXS* work discussed in Ref. (Izzo *et al.*, 2010) demonstrated that the Sachs–Teller theory (Sachs and Teller 1941) provides an accurate and consistent description of the spectral shape of a classical diatomic fluid.

According to the theoretical treatment by Krieger and Nelkin (Krieger and Nelkin, 1957), the spectrum at the crossover to the ST regime is isomorph to the one of a monatomic system approaching the IA limit, i.e. it has a Gaussian shape and its second moment is simply related to the mean single-molecule (roto-translational) kinetic energy, $\langle K.E.\rangle_{RT}$. Consistently, Izzo and collaborator used the additive approach in Eqs. ((7.1))–((7.4)) to model the *IXS* lineshape of liquid iodine in the proximity of the ST regime. The expected values of the model parameters were those predicted by the ST theory, which considers both rotational and translational degrees of freedom of molecules.

Due to the non-quantum nature of the sample considered, it is not surprising that the value of the mean roto-translational single-molecule kinetic energy extracted from the slope of the second central moment was found consistent with the value expected for a classical diatomic liquid ($5k_B T/2$) within a 4% uncertainty. Furthermore, the effective mass of the struck molecule experiencing a roto-translational recoil was found consistent with the Sachs–Teller prediction to within 1%. Overall, the main result achieved by this work was a robust consistency check for the *ST* description of the spectral shape measured by *DIXS* techniques on a simple molecular fluid.

Why are *DIXS* measurements relatively uncommon in the literature?

The substantial broadening of the sample spectrum at high Q's makes resolution requirements less stringent for *DIXS* measurements than for ordinary (lower-Q) *IXS* ones; this ultimately enhances the photon flux at the sample, partially compensating for the competing high-Q decrease of the *IXS* cross-section.

Furthermore, the virtual absence of kinematic limitations in principle enables a single *IXS* spectrometer to map the whole (Q, ω)-crossover between hydrodynamic and single-particle regimes. This advantage is crucial to circumvent the difficulties associated with the joint analysis of measurements performed by different instruments.

Given these grounds, one could wonder why *DIXS* measurements are still so sporadic in the literature, and the few works available are mostly limited to the scope of feasibility demonstrations. There are a few concomitant factors explaining this situation, including the stronger interest of *IXS* community toward the collective, rather than the single-particle dynamics of disordered systems. Furthermore, *DIXS* studies on quantum effects are held back by count rate limitations as these effects are best observable in low Z materials, such as D_2, H_2, and ^3He, and ^4He, which happen to be weak X-Ray scatterers. Finally, even the highest Q values reachable by

DIXS are still more than an order of magnitude below those covered by *DINS*.

7.10. Gaining insight from spectral moments: The onset of quantum effects

Lineshape models discussed thus far are intrinsically classical in character, as they can be derived by assuming that all relevant variables and operators are mutually commuting. Within this classical treatment, all correlation functions are even functions of time, and the corresponding spectra are symmetric in ω. It is generally assumed that deviations from this classical behavior only arise from the detailed balance principle introduced at the end of Chapter 6, which causes an asymmetry of the spectral profile.

When dealing with the microscopic behavior of many atoms systems, it is customary to assume that quantum deviations mainly split into two distinct categories:

(1) Diffraction or delocalisation effects, arising from the noncommutative nature of operators.

(2) Exchange effects, reflecting the symmetry restrictions in systems of identical particles, which ultimately determine the statistical population of sample states with $\hbar\omega$ energy.

Both these effects depend on the de Broglie wavelength $\lambda_B = h/\sqrt{2\pi M k_B T}$. However, diffraction effects become sizable when the interparticle potential varies appreciably over distances comparable with λ_B, while exchange effects come into play at very low temperatures only, through a few "spectacular" quantum-mechanical phenomena, such as Bose condensation and superfluidity, which are observable up to macroscopic length scales.

Here we only deal with diffraction effects which are dominating in the so-called moderately quantum fluids. On a general ground, quantum effects can be adequately observed under two conditions:

(1) Interatomic separations and "in cage" rattling periods should be comparable or shorter than λ_B and $T_B = \lambda_B/c$, respectively.

(2) The distance range, $d \subseteq (0, 2\pi/Q)$, and timescale, $t \subseteq (0, 2\pi/\omega)$, probed by the measurement should roughly match λ_B and T_B, respectively.

The first condition implies that the system "inherently" exhibits a quantum behavior, while the second guarantees that the measurement is tuned to observe it.

For quantum fluids at few tens of K, these conditions are met in a (Q, ω)-window exceeding by a few orders of magnitude the range covered by Brillouin light scattering, i.e. falling right in the mesoscopic domain of *IXS* and *INS*.

From the experimental side, a few *INS* studies of quantum effects based upon the determination of $S(Q, \omega)$ spectral moments in simple fluids are available in the literature; these include, for instance, two *INS* works on liquid Ne (van Well and de Graaf 1985) and Ar (van Well, P. Verkerk *et al.*, 1985). As an example, the results obtained on the argon sample are reported in Figure 7-14, which displays the Q-dependence of the first spectral moment normalized to the recoil frequency and the second spectral moment divided by its classical value. The plot includes for comparison the value obtained using the $\hbar^2$-truncated expansion, which, as shown in Appendix 7B, reads as:

$$\langle \omega^2 \rangle = \omega_r^2 + \frac{Q^2}{2M} \left[2k_B T + \frac{\hbar^2}{6k_B T} \left(\Omega_0^2 - \Omega_Q^2 \right) \right],$$

where the Einstein frequency Ω_0 and Ω_Q, respectively defined in Eqs. (2.21) and (6.12), were computed using a Lennard–Jones potential. From Figure 7-14, it emerges that the normalized first spectral moment is, apart from the lowest Q values, very close to its expected value, which demonstrates the ability of the measurement to provide a rigorous determination of the first moment. However, the situation is not as satisfactory for the normalized second spectral moment; although its absolute values only slightly exceed the $\hbar^2$-truncated theoretical prediction, their actual Q-dependence is likely concealed by significant statistical oscillations.

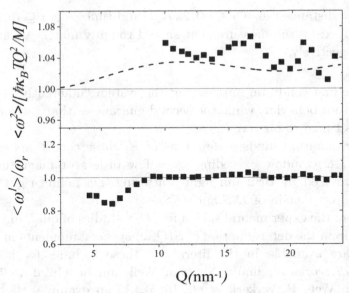

Figure 7-14:   Upper panel — the normalized second spectral moment of the *INS* spectra of liquid Ar at $T = 120\,\text{K}$ and $P = 11.5$ bar is reported as a function of Q. Data are normalized to their classical value (squares) and compared with the theoretical prediction, obtained using a Lennard–Jones model potential (see text). Lower panel — the corresponding first spectral moment normalized to its exact value ω_r (squares). In both panels, the horizontal solid line represents the unit value for reference. Data in the upper and lower panels are redrawn from Refs. (van Well and de Graaf 1985) and (van Well, P. Verkerk *et al.*, 1985), respectively.

As far as available *IXS* results are concerned, investigations of quantum effects on spectral moments mainly rest on the integration of analytical lineshape models, rather than measured spectral intensities, as discussed in the next section.

7.11.   *IXS* studies of quantum effects in simple liquids

Direct evaluations of spectral moments from the integrated scattering signal are hampered by a number of spurious intensity effects, as well as by the nontrivial contribution from the energy resolution profile.

Spurious intensity contributions can be eliminated by dealing with $\langle \omega^n \rangle / \langle \omega^1 \rangle$ ratios, as they equally affect numerator and denominator. This strategy is particularly convenient, as the analytical expression of the first spectral moment is exactly known and straightforward for a monatomic system (see Appendix 7B). Conversely, the correction for the resolution contribution would require a numerical deconvolution of the measured lineshape, which yields uncertain results when the scattering signal has sizable inelastic features. This being the case, the resolution-free signal can be estimated through a modeling of the lineshape.

The *IXS* studies of quantum effects on spectral moments available in the literature were performed on supercritical He (Verbeni, Cunsolo *et al.*, 2001) and liquid Ne (Monaco, Cunsolo *et al.*, 2002, Cunsolo, Monaco *et al.*, 2003). In these studies, spectral moments were computed from the integration of unconvoluted model lineshapes, as estimated through best-fit values of $I(Q, \omega)$ in Eq. (6.22).

The first *IXS* work investigated quantum effects on the spectrum of supercritical helium at moderate Q's by varying the thermodynamic conditions of the sample, while the two successive ones were performed on a liquid Ne sample at fixed thermodynamic conditions, while spanning the broad Q-crossover from the classical (continuous) to the quantum (single particle) regimes.

A critical variable in the focus of these works was the reduced second moment:

$$R_2 = \frac{\langle \omega^2 \rangle}{\langle \omega^1 \rangle} - \omega_r. \tag{7.5}$$

The two opposite, classical and quantum, limits of this variable are met upon integrating the fluid's response over long and short distances, and are respectively given by:

$$\lim_{Q \to 0} R_2(Q) = \frac{4}{3} \langle K.E. \rangle_C = 2k_B T$$

$$\lim_{Q \to \infty} R_2(Q) = \frac{4}{3} \langle K.E. \rangle,$$

where it was considered that, for a monatomic system, the classical value of the mean single-particle kinetic energy is $\langle K.E. \rangle_C = 3/2K_B T$.

The above formulas suggest that, by mapping a sufficiently large Q-range, a single spectroscopic measurement can in principle cover the whole crossover from the classic to the quantum regime. Furthermore, from the measurement of the second spectral moment, one should be able to achieve a straightforward measurement of $\langle K.E.\rangle$ and estimate the relative weight of quantum deviation as $(\langle K.E.\rangle - \langle K.E.\rangle_C)/\langle K.E.\rangle_C$.

Figure 7-15 provides a general idea on the experimental values of $R_2(Q)$ obtained for two simple monatomic fluids, He and Ne, in various thermodynamic conditions indicated in the plot. Plotted values are taken from Refs. (Verbeni, Cunsolo *et al.*, 2001) and (Cunsolo, 1999) respectively, after normalization to the classical value. It is remarkable that the substantially larger amplitude of quantum deviations in He data is due to its generally large de Broglie wavelength. This is always the case except at the highest temperature $(T = 294\,\mathrm{K})$ where He also behaves as a classical fluid.

Furthermore, the plot shows that the classical value is reached by both systems in the macroscopic, $Q = 0$, limit, while at higher Q's, that is over shorter distance ranges, quantum effects clearly show up, especially at lower T and larger number density n. Indeed, in these thermodynamic states, the mean free path more closely approaches the de Broglie wavelength, both being in the Q^{-1}-range.

Among various *IXS* results in this area, it is also worth mentioning two joint *IXS* measurements and quantum Path Integral Monte Carlo (*PIMC*) simulations (Cunsolo *et al.*, 2005, Cunsolo *et al.*, 2002, Pratesi *et al.*, 2002) on H_2 and D_2, in corresponding thermodynamic states, i.e. states having the same reduced thermodynamic coordinates T/T_c and n/n_c, where, the index "c" labels the value at the critical point. In fact, since classical fluids in corresponding states should exhibit the same microscopic response (Guggenheim, 1945), differences possibly observed in the scattering profiles from the two samples would have revealed the onset of quantum effects, supposedly more pronounced in the system with larger de Broglie wavelength, i.e. H_2. Indeed Figure 7-16 confirms this hypothesis by comparing the *IXS* spectral shapes of H_2 and D_2 measured in Ref. (Cunsolo, Colognesi *et al.*, 2005) at $Q = 12.8\,\mathrm{nm}^{-1}$. It appears

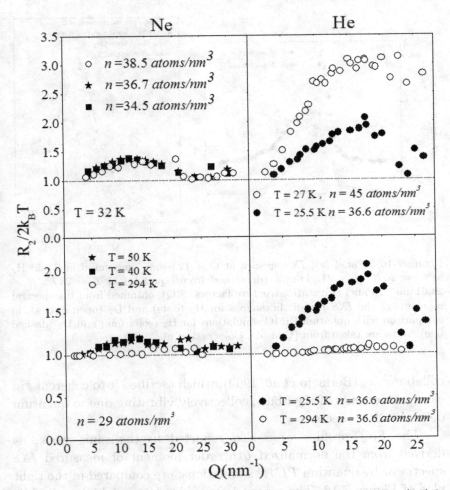

Figure 7-15: The Q-dependence of the spectral moment ratio in Eq. (7.5) in supercritical He and liquid and supercritical Ne at the indicated thermodynamic conditions and normalized to the classical value $2k_BT$. The normalized, (semi-classical) unit value is also shown as a dashed horizontal line for reference. Data are from Refs. (Verbeni, Cunsolo *et al.*, 2001) and (Cunsolo, 1999).

that the spectrum of liquid hydrogen bears evidence for a more pronounced inelastic shoulder on the Stokes side, indicating the presence of a long-living collective excitation. A similar spectral feature was also reported by an earlier *INS* work of Bermejo and

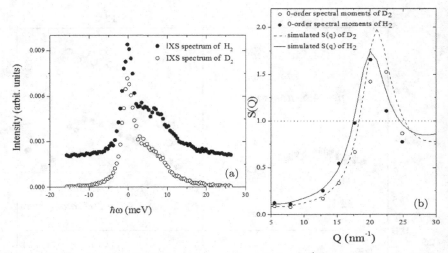

Figure 7-16:   Panel (a): *IXS* spectra at $Q = 12.8\,\text{nm}^{-1}$ measured in liquid H_2 at $T = 20\,K$, $n = 21.24\,\text{nm}^{-3}$ (dots) and from liquid D_2 at $T = 23\,K$, $N = 24.61\,\text{nm}^{-3}$. Panel (b): static structure factors, S(Q), obtained from the spectral moments of the *IXS* best-fit lineshapes for H_2 (dots) and D_2 (open circles), in comparison with quantum PIMC simulations for H_2 (solid line) and D_2 (dashed line). Data are taken from (Cunsolo, Colognesi *et al.*, 2005).

collaborators (Bermejo *et al.*, 1999), which ascribed it to coherent "in cage" movements of H_2 atoms, collectively vibrating due to quantum delocalisation effects.

The $S(Q)$-profiles were also reported in the same work, as derived from the normalized θth order moment of measured *IXS* spectra or by quantum *PIMC* simulations, are compared in the right plot of Figure 7-16. The comparison evidences a slight low-Q shift of the dominant H_2 diffraction maximum as compared to its D_2 counterpart.

This shift is caused by the higher quantum delocalisation of H_2 molecules' centroids, which fosters their intrusion into the repulsive region of the first neighbour interaction potential. These more extended visits into a repulsive environment cause a rebound of the single-molecule ultimately resulting in an increased first-neighbour distance, which slightly shifts the first $S(Q)$ maximum to lower Q's.

A potentially controversial aspect of the analysis discussed thus far is that quantum effects are sought for in the best-fit model lineshapes, rather than the raw scattering profiles. This strategy might deliver model-dependent results; also, the lineshape model considered in these studies is semi-classical, as it assumes that quantum effects only influence the population of ω-states through the detailed balance principle (see last section of chapter 6).

More specifically, the "classical" (symmetrical) lineshape was described with a simple viscoelastic lineshape model, which yields converging spectral moments up to the fourth-order. It was also recognized (Rabani and Reichman, 2002) that the presence of mild to moderate quantum deviations is compatible with the choice of a single exponential *ansatz* for the memory function, and a "quantum viscoelastic model" can be used to formally account for quantum deviations. In moderately quantum fluids these effects mainly result in a global softening of molecular interactions stemming from the quantum delocalization which induces higher fluctuations in the atomic position and an overall loss of rigidity of the system.

In recent times, *IXS* studies of quantum deviations based on the analysis of spectral moments reached some impasse, partly owing to the circumstance that a more rigorous assessment of these effects would require direct integration of the scattering signals rather than lineshape modeling. This endeavour would critically benefit from an overall narrowing and a tail suppression of current *IXS* resolutions, both achievements minimizing the instrumental contribution to spectral moments.

Appendix 7A: Brief hints on the Enskog theory formalism (Kamgar-Parsi *et al.*, 1987)

As discussed in previous sections, the general aim of the KT approach is to derive the correlation matrix associated with a generic fluctuating variable $\alpha_j(Q,t)$ of the jth atom, which for an N-atoms system, reads as:

$$\mathbf{C}_{jk}(Q,t) = \langle \alpha_k^\star(Q) \exp\left(tL_{HS}^N\right) \alpha_j(Q) \rangle$$

where:

$$L_{\text{HS}}^{\text{N}} = \sum_{j=1}^{N} \boldsymbol{v}_j \cdot \frac{\partial}{\partial \boldsymbol{r}_j} + \sum_{\substack{j \neq k=1 \\ j<k}}^{N} \boldsymbol{T}_{jk} \tag{7A.1}$$

while $\boldsymbol{T}_{jk}$ is an operator accounting for inter-particle collisions.

We can now introduce the first three normalized phase-space variables for the system:

$$\alpha_1(Q) = \frac{1}{\sqrt{N}} \sum_{j=1}^{N} \phi_1(\mathbf{v}_j) \exp(i\mathbf{Q} \cdot \mathbf{R}_j) \tag{7A.2}$$

$$\alpha_2(Q) = \frac{1}{\sqrt{N}} \sum_{j=1}^{N} \phi_2(\mathbf{v}_j) \exp(i\mathbf{Q} \cdot \mathbf{R}_j) \tag{7A.3}$$

$$\alpha_3(Q) = \frac{1}{\sqrt{N}} \sum_{j=1}^{N} \phi_3(\mathbf{v}_j) \exp(i\mathbf{Q} \cdot \mathbf{R}_j) \tag{7A.4}$$

here $\mathbf{r}_j$ and $\mathbf{v}_j$ are the $t = 0$ position and velocity of the jth hard sphere. The fluctuating variables defined above correspond to the $t = 0$ fluctuations of the the density of mass, longitudinal momentum and energy, respectively. The coefficients $\varphi_j(\boldsymbol{v})$ in the above equations are functions of the single particle velocity, $\boldsymbol{v}$, which read as:

$$\phi_1(\mathbf{v}_j) = \frac{1}{\sqrt{S(Q)}}$$

$$\phi_2(\mathbf{v}_j) = \sqrt{\frac{M}{k_B T}} \frac{\mathbf{Q} \cdot \mathbf{v}}{Q}$$

$$\phi_3(\mathbf{v}_j) = \frac{1}{\sqrt{6}} \left(3 - \frac{Mv^2}{k_B T} \right)$$

Notice that both N-particle variables $\alpha_j(Q)$ and the one-particle ones $\phi_j(\boldsymbol{v})$ form orthonormal sets, i.e.:

$$\langle \alpha_j \,() \, \alpha_j^{\star}(Q) \rangle = \langle \varphi_j(\mathbf{v}_1) \varphi_k^{*}(\mathbf{v}_1) \rangle_1 = \delta_{jk}, \tag{7A.5}$$

where the suffixes "N" and "1" on the right angle brackets label, respectively, the ordinary N-particle thermal average performed on the system at equilibrium and single particle or self average. The latter is defined, for any function $h(\mathbf{v}_1)$, as follows:

$$\langle h(\mathbf{v}_1)\rangle_1 = \int d\mathbf{v}_1 \Phi(\mathbf{v}_1) h(\mathbf{v}_1), \qquad (7A.6)$$

where we have introduced the normalized Boltzmann distribution:

$$\Phi(\mathrm{v}) = \left(\frac{M}{2\pi k_B T}\right)^{\frac{3}{2}} \exp\left(-\frac{M v^2}{2 k_B T}\right), \qquad (7A.7)$$

The core idea of the revised Enskog theory (RET) is to replace the N-particle variables in Eqs. (7A.2-4) by the corresponding single-particle variables and to represent the dynamic response of the fluid through the single-particle correlations of single-particle variables in Eqs. (7A.5-7). Therefore, the jk element of the correlation matrix reads as:

$$C_{jk}^{\mathrm{E}}(Q,t) = \langle \varphi_j^*(\mathbf{v}_1)\exp[t L_{\mathrm{E}}(\mathbf{Q},\mathbf{v}_1)]\,\varphi_k(\mathbf{v}_1)\rangle_1. \qquad (7A.8)$$

The time-propagator in the above formula contains the generalized single-particle Enskog operator, which for a hard-sphere system can be represented as (de Schepper and Cohen, 1980):

$$L_{\mathrm{E}}(\mathbf{Q},\mathbf{v}_1) = -i\mathbf{Q}\cdot\mathbf{v}_1 + n g(\sigma_{HS})\mathbf{\Lambda}_Q + n\mathbf{A}_{\mathrm{Q}}, \qquad (7A.9)$$

which essentially replaces the N-particle operator $\mathbf{L}_{\mathrm{HS}}^{\mathrm{N}}$ in Eq. (7A.1). Here, $g(\sigma_{HS})$ is the pair distribution function corresponding to hard-spheres of diameter σ_{HS} at contact. The first term is the Fourier transform of the free streaming term. The second term accounts for binary collision, also included in the original Boltzmann equation; it acts on a generic function $h(\mathbf{v}_1)$ as:

$$\Lambda_Q h(\mathbf{v}_1) = -\sigma_{HS}^2 \int d\hat{\sigma}_{HS} \int d\mathbf{v}_2 \Phi(\mathbf{v}_2)\,\delta\theta(\delta)$$
$$\times \left\{ h(\mathbf{v}_1) - h(\mathbf{v}_1') + \exp(-i\mathbf{Q}\cdot\hat{\boldsymbol{\sigma}}_{HS})\left[h(\mathbf{v}_2) - h(\mathbf{v}_2')\right]\right\},$$

where $\hat{\sigma}_{HS}$ is a unit vector which defines the collision geometry through the vector $\boldsymbol{\sigma}_{HS} = \sigma_{HS}\hat{\sigma}_{HS}$ connecting the centres of the colliding hard spheres, $\theta(\delta)$ is the Heaviside step function of the variable $\delta = (\boldsymbol{v}_1 - \boldsymbol{v}_2)\cdot\hat{\sigma}_{HS}$, while $\boldsymbol{v}'_{1,2} = \boldsymbol{v}_{1,2} \mp \delta\hat{\sigma}_{HS}$ are the velocities the hard spheres "1" and "2" where the presence of the prime distinguishes the post-collision values from the pre-collision ones.

The third term of the Enskog operator has no analogue in the original Boltzmann equation, and it contains the static structure factor $S(Q)$, i.e. the "mean field" in which the binary collisions take place, and it acts on a function $h(\boldsymbol{v}_1)$ as follows:

$$\boldsymbol{A}_Q h\left(\boldsymbol{v}_1\right) = (1/\rho)\left[C(Q) - g\left(\sigma_{HS}\right)C_0(Q)\right]\int d\boldsymbol{v}_2 \Phi\left(\boldsymbol{v}_2\right) i\boldsymbol{Q}\cdot\boldsymbol{v}_2 h\left(\boldsymbol{v}_2\right)$$

where $C(Q) = 1 - 1/\S(Q)$ and $C_0(Q) = \lim_{\rho\to 0}C(Q)$. At this stage, one can write the Enskog operator in the basis of the eigenmodes, where it assumes the following diagonal form:

$$L_E\left(\boldsymbol{Q},\boldsymbol{v}_1\right) = -\sum_h |\phi_h\left(\boldsymbol{Q},\boldsymbol{v}_1\right)\rangle \lambda_h(Q)\langle\psi_h\left(\boldsymbol{Q},\boldsymbol{v}_1\right)|.$$

Here $|\phi_h\left(\boldsymbol{Q},\boldsymbol{v}\right)\rangle$, $\langle\psi_h\left(\boldsymbol{Q},\boldsymbol{v}\right)|$ and $\lambda_h(Q)$ are the hth left and right eigenfunction and associated eigenvalue of the Enskog operator, respectively. In the eigenmodes' basis, the dynamic structure factor can be simply written as:

$$S(Q,\omega) = [S(Q)/\pi]\, Re\left\langle\frac{1}{i\omega - L_E\left(\boldsymbol{Q},\boldsymbol{v}_1\right)}\right\rangle_1$$

$$= [S(Q)/\pi]\, Re\left\{\sum_j \frac{A_j(Q)}{i\omega + \lambda_j(Q)}\right\},$$

where $A_j(Q) = \langle\psi_j(\boldsymbol{Q},\boldsymbol{v}_1)\rangle_1\langle\psi_j^*(\boldsymbol{Q},\boldsymbol{v}_1)\rangle_1$ is the complex amplitude of the jth eigenmode. In general, the eigenvalues $\lambda_h(Q)$ can either be real or appear in complex conjugate pairs, its imaginary and real parts representing the damping and the propagating frequency of the hth mode, respectively.

The above formula expresses the spectrum as a sum of (complex) Lorentzian terms. In general, this kind of expansion is a quite general method to represent the spectrum. In particular, if the chosen fluctuating variables are the hydrodynamic variables, the resulting spectrum reproduces the linear hydrodynamic result, yielding a spectral shape analytically isomorph to the Brillouin triplet, that is the *GH* model profile in Eq. (6.6).

To bestow the model with a deeper physical ground, one can superimpose to it sum rules of increasing order n, which leads to:

$$\sum_j A_j(Q)[\lambda_j(Q)]^n = R_n(Q), \qquad (7A.10)$$

where, for instance $R_0(Q) = \sum_j A_j(Q) = S(Q)$, $R_2(Q) = \sum_j A_j(Q)\lambda_j^2(Q) = k_B T Q^2/M$ and so forth, as can be deduced from the short time expansion of the intermediate scattering function.

The eigenvalues of the generalized Enskog operator $L_E(\boldsymbol{Q}, \mathbf{v})$ were computed in the mid-1970s, by Kamgar–Parsi and collaborators for a hard-sphere system (Kamgar–Parsi *et al.*, 1987). This study led to identifying several density and Q regimes. Different values of V/V_0 — with V and V_0 the specific and the close packing volumes, respectively — characterize the density regimes. Namely:

(1) A low-density regime which corresponds to $V/V_0 < 0.1$.
(2) An intermediate-density regime for which $0.1 < V/V_0 < 0.35$.
(3) A dense-fluid regime $0.35 < V/V_0 < 0.70$.

The various Q-regimes depend on how Q compares with the inverse of the Enskog mean free path $l_E = 1/\left[\sqrt{2}\pi n g\,(\sigma_{HS})\,\sigma_{HS}^2\right]$. These regimes are:

(1) $0 < Ql_E < 0.1$: this regime is the kinetic analogue of hydrodynamical regime characterized by three (heat and sound) modes;
(2) $0.1 < Ql_E < 1$: this regime is the finite-Q extension of the hydrodynamic regime in which generalized hydrodynamic modes plus two kinetic eigenmodes of $L_E(\boldsymbol{Q}, \mathbf{v})$ are expected to appear in the spectrum. Usually, "kinetic" eigenmodes are distinguished from hydrodynamic ones as their dominant frequency does not vanish in the low wavevector limit;

(3) $1 < Ql_E < 3$: this is a new dynamic regime where all $L_E(\mathbf{Q}, \mathbf{v})$ eigenmodes play an essential role in the spectrum of density fluctuations, and no simplified description exists;

(4) $3 < Ql_E < \infty$: in this regime the eigenmodes associated with the individual-particle (free) streaming are dominating.

Appendix 7B: Handling quantum effects analytically (Fredrikze, 1983)

As previously discussed, quantum deviations in the density correlation function mainly emerge at short times. Therefore, their signature can be sought for in the coefficients governing the short-time behavior of the intermediate scattering function, $F(q, t)$, *i.e.* n-order spectral moments:

$$\langle \omega^n \rangle = i^{-n} \frac{d^n}{dt^n} F(q, t) \Big|_{t=0}$$

$$= i^{-n} \frac{1}{N} \sum_{j,k=1}^{N} \left\langle \exp\left(-i\mathbf{Q} \cdot \mathbf{R}_k\right) \frac{d^n}{dt^n} \{\exp\left[i\mathbf{Q} \cdot \mathbf{R}_j(t)\right]\}_{t=0} \right\rangle,$$

with the usual shorthand notation $\mathbf{R}_k = \mathbf{R}_k(0)$. When handling the nth spectral moments explicitly it is useful to consider that, for a variable A which does not depend on time explicitly, the n-order derivative can be computed by applying n times the Heisenberg equation:

$$\frac{d^n A}{dt^n} = \left(\frac{1}{\hbar}\right)^n [\ldots [[A, H], H] \ldots, H]$$

where $[A, B]$ is the commutator of A and B and H the Hamiltonian of the system. when considering the spectral moment we thus can write:

$$\langle \omega^n \rangle = \left(\frac{-1}{\hbar}\right)^n$$

$$\times \frac{1}{N} \sum_{j,k=1}^{N} \langle \exp\left(-i\mathbf{Q} \cdot \mathbf{R}_k\right) [\ldots [[\exp\left(i\mathbf{Q} \cdot \mathbf{R}_j\right), H], H] \ldots, H] \rangle$$

Here the Hamiltonian, H, of a system of N atoms interacting through an additive and pairwise potential reads as:

$$H = -\sum_i \frac{\hbar^2 \nabla_i^2}{2M} + \sum_{j<k} V\left(|\mathbf{R}_{jk}|\right),$$

with $\boldsymbol{R}_{jk} = \boldsymbol{R}_j - \boldsymbol{R}_k$ being the vector connecting the jth and kth atoms,

Evaluating the commutator explicitly and using some time-reversal properties of involved operators, one can derive the following expression for the first spectral moment:

$$\langle \omega^1 \rangle = \omega_r.$$

This result is quite general and can be derived computation of higher-order momentum can be worked out following a similar procedure. For instance, the following expression can be derived for the second spectral moment (see, e.g. Stringari 1992):

$$\langle \omega^2 \rangle = \omega_r^2 \left[2 - S(Q)\right] + \frac{\omega_r}{\hbar} \left[\frac{4}{3} \langle K.E. \rangle + K(Q)\right]$$

The above expression contains the mean single-particle kinetic energy:

$$\langle K.E. \rangle = \frac{3}{N} \sum_{j=1}^{N} \left\langle \frac{\left(p_j^Q\right)^2}{2M} \right\rangle,$$

with $p_i^Q = i\hbar \mathbf{Q} \cdot \nabla_i$ being the projection of the momentum of the ith atom along $\mathbf{Q}$; furthermore, we have introduced the so-called kinetic factor:

$$K(Q) = \frac{1}{NM} \sum_{j\neq k=1}^{N} \left\langle \cos\left(\mathbf{Q} \cdot \mathbf{r}_{jk}\right) p_j^Q p_k^Q \right\rangle.$$

Notice that, in the classical limit, one has $\langle \omega^2 \rangle = 4/3\,(\omega_r/\hbar)$ $\langle K.E. \rangle_c = Q^2 k_B T/M$, as readily follows from recognising that, for a classical monatomic fluid, $\langle K.E. \rangle_c = 3k_B T/2$.

For moderately quantum particles interacting through an additive and pairwise interaction potential, a $\hbar$-expansion of the second spectral moment can be explicitly performed (Fredrikze, 1983). To the $\hbar^2$-order, it yields the approximated formula:

$$\langle \omega^2 \rangle \approx \omega_r^2 + \frac{\omega_r}{\hbar} \left[2k_B T + \frac{\hbar^2}{6k_B T} \left(\Omega_0^2 - \Omega_Q^2 \right) \right].$$

which relates $\langle \omega^2 \rangle$ to the Einstein frequency (see Eq. 2.21) and the Q-dependent factor introduced in Eq. (6.12).

The third spectral moment can be computed with the same procedure used for lower-order moments. For an additive pair potential, this calculation leads to the following result (Puff, 1965):

$$\langle \omega^3 \rangle = \omega_r^3 + 4\frac{\omega_r^2}{\hbar} \langle K.E. \rangle + \left(\Omega_0^2 - \Omega_Q^2 \right).$$

References

Bafile, U., Verkerk, P., Barocchi, F., de Graaf, L. A., Suck, J. & Mutka, H. 1990. *Phys. Rev. Lett.*, **65**, 2394.

Balucani, U., Ruocco, G., Torcini, A. & Vallauri, R. 1993. *Phys. Rev. E*, **47**, 1677.

Balucani, U., Vallauri, R. & Gaskell, T. 1987. *Phys. Rev. A*, **35**, 4263.

Bell, H., Moellerwenghoffer, H., Kollmar, A., Stockmeyer, R., Springer, T. & Stiller, H. 1975. *Phys. Rev. A*, **11**, 316.

Bell, H. G., Kollmar, A., Alefeld, B. & Springer, T. 1973. *Phys. Lett. A*, **45**, 479.

Bermejo, F. J., Fåk, B., Bennington, S. M., Fernández-Perea, R., Cabrillo, C., Dawidowski, J., Fernández-Diaz, M. T. & Verkerk, P. 1999. *Phys. Rev. B*, **60**, 15154.

Bodensteiner, T., Morkel, C., Gläser, W. & Dorner, B. 1992. *Phys. Rev. A*, **45**, 5709.

Campa, A. & Cohen, E. G. 1989. *Phys. Rev. A*, **39**, 4909.

Canales, M. & Padró, J. A. 1999. *Phys. Rev. E*, **60**, 551.

Castillo, R. & Castañeda, S. 1988. *Int. J. Thermophys.*, **9**, 383.

Chen, S. H., Eder, O. J., Egelstaff, P. A., Haywood, B. C. G. & Webb, F. J. 1965. *Phys. Lett.*, **19**, 269.

Cohen, E. G. D. 1993. *Physica A*, **194**, 229.

Cohen, M. L. 1986. *Science*, **234**, 549.

Copley, J. & Rowe, J. 1974. *Phys. Rev. Lett.*, **32**, 49.

Cunsolo, A. 1999. Thesis, Universite' J. Fourier.
Cunsolo, A., Colognesi, D., Sampoli, M., Senesi, R. & Verbeni, R. 2005. *J. Chem. Phys.*, **123**, 114509.
Cunsolo, A., Monaco, G., Nardone, M., Pratesi, G. & Verbeni, R. 2003. *Phys. Rev. B*, **67**.
Cunsolo, A., Pratesi, G., Colognesi, D., Verbeni, R., Sampoli, M., Sette, F., Ruocco, G., Senesi, R., Krisch, M. H. & Nardone, M. 2002. *J. Low Temp. Phys.*, **129**, 117.
Cunsolo, A., Pratesi, G., Ruocco, G., Sampoli, M., Sette, F., Verbeni, R., Barocchi, F., Krisch, M., Masciovecchio, C. & Nardone, M. 1998. *Phys. Rev. Lett.*, **80**, 3515.
de Schepper, I. M. & Cohen, E. G. D. 1980. *Phys. Rev. A*, **22**, 287.
de Schepper, I. M., van Rijs, J. C., van Well, A. A., Verkerk, P., de Graaf, L. A. & Bruin, C. 1984a. *Phys. Rev. A*, **29**, 1602.
de Schepper, I. M., Verkerk, P., van Well, A. A. & de Graaf, L. A. 1984b. *Phys. Lett. A*, **104**, 29.
Dorner, B., Plesser, T. & Stiller, H. 1967. *Discuss. Faraday Soc.*, **43**, 160.
Egelstaff, P. A., Glaser, W., Litchinsky, D., Schneider, E. & Suck, J. 1983. *Phys. Rev. A*, **27**, 1106.
Egelstaff, P. A. 1967. *An Introduction to the Liquid State*, London, New York, Academic Press.
Ernst, M. H., Haines, L. K. & Dorfman, J. R. 1969. *Rev. Mod. Phys.*, **41**, 296.
Glyde, H. R. 1994. *Phys. Rev. B*, **50**, 6726.
Guggenheim, E. A. 1945. *J. Chem. Phys.*, **13**, 253.
Henshaw, D. G. & Woods, A. D. B. 1961. *Phys. Rev.*, **121**, 1266.
Izzo, M. G., Bencivenga, F., Cunsolo, A., Di Fonzo, S., Verbeni, R. & Gimenez De Lorenzo, R. 2010. *J. Chem. Phys.*, **133**, 124514.
Kambayashi, S. & Hiwatari, Y. 1994. *Phys. Rev. E*, **49**, 1251.
Kamgar-Parsi, B., Cohen, E. G. & de Schepper, I. M. 1987. *Phys. Rev. A*, **35**, 4781.
Krieger, T. J. & Nelkin, M. S. 1957. *Phys. Rev.*, **106**, 290.
Kroô, N., Borgonovi, G., Sköld, K. & Larsson, K.-E. 1964. Inelastic Scattering of Cold Neutrons by Condensed Argon. SYMPOSIUM ON INELASTIC SCATTERING OF NEUTRONS, Bombay. IAEA.
Monaco, G., Cunsolo, A., Pratesi, G., Sette, F. & Verbeni, R. 2002. *Phys. Rev. Lett.*, **88**, 227401.
Mountain, R. D. 1982. *Phys. Rev. A*, **26**, 2859.
NIST. *Thermodynamic data are from the database:* http://webbook.nist.gov/chemistry/form-ser.html.
Ohse, R. W. 1985. *Handbook of Thermodynamic and Transport Properties of Alkali Metals*, Blackwell Oxford.

Postol, T. A. & Pelizzari, C. A. 1978. *Phys. Rev. A*, **18**, 2321.

Pratesi, G., Colognesi, D., Cunsolo, A., Verbeni, R., Nardone, M., Ruocco, G. & Sette, F. 2002. *Philos Mag B*, **82**, 305.

Rabani, E. & Reichman, D. R. 2002. *J. Chem. Phys.*, **116**, 6271.

Randolph, P. D. & Singwi, K. S. 1966. *Phys. Rev.*, **152**, 99.

Scopigno, T., Balucani, U., Cunsolo, A., Masciovecchio, C., Ruocco, G., Sette, F. & Verbeni, R. 2000. *Europhys. Lett.*, **50**, 189.

Senesi, R., Andreani, C., Colognesi, D., Cunsolo, A. & Nardone, M. 2001. *Phys. Rev. Lett.*, **86**, 4584.

Silver, R. N. & Sokol, P. E. 2013. *Momentum Distributions*, Springer Science & Business Media.

Skold, K. & Larsson, K. E. 1967. *Phys. Rev.*, **161**, 102.

Stringari, S., Pitaevskii, L., Stamper-Kurn, D. M. & Zambelli, F. 2002. Momentum Distribution of a Bose Condensed Trapped Gas. *In*: Martellucci, S., Chester, A. N., Aspect, A. & Inguscio, M. (eds.) *Bose-Einstein Condensates and Atom Lasers*. Boston, MA: Springer US.

van Well, A. A. & de Graaf, L. A. 1985. *Phys. Rev. A*, **32**, 2396.

van Well, A. A., Verkerk, P., de Graaf, L. A., Suck, J. B. & Copley, J. R. D. 1985. *Phys. Rev. A*, **31**, 3391.

Verkerk, P. & van Well, A. A. 1986. *Physica B+C*, **136**, 168.

Yip, S. 1979. *Annu. Rev. Phys. Chem.*, **30**, 547.

Chapter 8

Terahertz relaxation phenomena in simple systems probed by *IXS*

8.1. Introductory topics

In the previous chapter we mentioned that a typical example of collective mode probed by a spectroscopic experiment is an acoustic wave propagating with a wavevector amplitude Q and angular frequency $\omega_s(Q)$. This wave can be portrayed as a chain of compression-rarefaction zones (*CRZ*s) oscillating with a period $T_{AW}(Q) = 2\pi/\omega_s(Q)$ and a wavelength $\lambda_{AW} = 2\pi/Q$. In these zones, the pressure is either higher or lower than the equilibrium value, and this causes time-dependent fluctuations of the local density.

During a compression half-cycle, an energy excess is stored and successively redistributed toward some internal degree of freedom having a timescale $\tau(Q)$. This energy rearrangement, which eventually leads to the recovery of local equilibrium, is accomplished provided there is a rough matching between $T_{AW}(Q)$ and $\tau(Q)$.

The mere existence of the characteristic timescale $\tau(Q)$, naturally splits acoustic propagation into the following two regimes:

(1) **Viscous regime**
    When *CRZ*s oscillate very slowly, *i.e.* $T_{AW} \gg \tau(Q)$, energy rearrangements are perceived by the sound wave as "instantaneous", and sound propagation occurs over successive equilibrium states of the system.

215

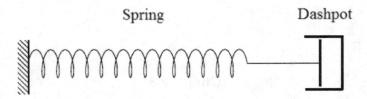

Figure 8-1:   The spring-dashpot model of viscoelasticity.

(2) **Elastic regime**

When, conversely, CRZs oscillate very rapidly, i.e. $T_{AW} \ll \tau(Q)$, internal degrees of freedom of the system are too slow to efficiently dissipate the energy of the acoustic wave, which mainly propagates without energy losses, *i.e.* elastically.

For acoustic periods intermediate between the elastic and viscous values, i.e. when $T_{AW} \approx \tau(Q)$, the response of the fluid combines both viscous and elastic aspects (**viscoelastic regime**).

Notice that this simplified scenario rests on the assumption that the relaxation process can be consistently described by a single timescale $\tau(Q)$, which is equivalent to assuming that an acoustic wave in the elastic regime essentially travels "unperturbed". Further below in this chapter we will consider a more complex picture of sound propagation in the "elastic" regime.

In the macroscopic ($Q = 0$) Maxwell's treatment of viscoelasticity the coexistence of viscous and elastic aspects is schematized with a two-element model sytem consisting of a linear spring and a linear viscous dashpot connected in series (see Figure 8-1), respectively embodying elastic and viscous aspects of a viscoelastic response.

On a general ground, the viscous limit is paradigmatic of the broad class of phenomena customarily referred to as "adiabatic". These are dynamical processes dominated by a slow and a fast timescale, respectively characterizing the time-dependent perturbation induced to the system and the response of the system to it.

Many physical problems fit into this category; a famous example we already encountered in this book is the so-called Born–Oppenheimer approximation, in which the electron and the nucleus dynamics can be identified as the rapidly responding system and the slowly dependent time perturbation, respectively. As discussed in Chapter 3, within this adiabatic approximation, the electron follows the atomic motion as a "slow drift" of the equilibrium centroid of its orbit.

Similarly, in the viscous regime, the rapidly relaxing system follows the acoustic perturbation which slowly moves its local equilibrium. Due to the infinitely rapid response of the system, the acoustic propagation occurs over successive equilibrium states.

Conversely, the "elastic" regime is a typical "anti-adiabatic" scenario, in which the perturbing density fluctuations become too rapid to be followed by the slow response of the system, which cannot restore equilibrium efficiently within a time comparable to the acoustic period. Consequently, in the elastic regime, the acoustic wave travels across successive non-equilibrium states of the fluid.

The coupling of density fluctuations with a relaxation process causes a breach of the adiabatic approximation, which, as discussed in Chapter 6, can be accounted for by including a non-Markovian, exponential decay of the memory function.

It is essential to recognize that the "viscoelasticity" is not an intrinsic property of some systems, but rather an effect observed whenever the frequency of the perturbation, or, equivalently, the one of the spectroscopic probe, roughly matches the inverse of the relaxation time, as schematically represented in Figure 8-2.

This concept is illustrated in the table below, which schematically exemplifies the response expected for three different prototypical samples when probed by methods covering different frequency windows.

The table also partially relates to one of the historical merits of *IXS*, i.e. the opportunity of probing picosecond relaxation processes at mesoscopic distances and without the kinematic limitations hampering *INS*. These relaxation phenomena are dominating in simple, low-viscosity systems such as noble gases, molten metals and alcohols.

It is perhaps interesting to compare the viscoelastic response to the case of a resistor-capacitor, or *RC*, electric circuit fed by a periodic electrical potential $V(t)$, which can have, for instance, the form of a square wave. In this analogy, $V(t)$ mimics the effect of the

System $\longrightarrow$	Supercooled glass former ($\tau \sim 10^{-6}$ s)	Glass former at melting ($\tau \sim 10^{-9}$ s)	Dense noble gas ($\tau \sim 10^{-12}$ s)
Technique $\downarrow$			
Ultrasound Spectroscopy ($\omega \sim$ KHz-MHz)	Viscoelastic	Viscous	Viscous
Brillouin Light Scattering ($\omega \sim$ GHz)	Elastic	Viscoelastic	Viscous
Inelastic X-Ray Scattering ($\omega \sim$ THz)	Elastic	Elastic	Viscoelastic

Figure 8-2:   A schematic illustration of how the viscoelasticity depends on the frequency of the probe and the relaxation time of the system.

acoustic wave, as schematized in Figure 8-3. This time-dependent perturbation stimulates a response, measurable as the difference of potential, $A(t)$, at the extremities of the capacitive element.

As known from Physics textbooks, such a response has the form of an exponential time-decay with time constant equal to the product between the resistance R and the capacity C of the electric circuit, *i.e.* $\tau_{RC} = RC$, which, in this analogy with viscoelastic phenomena, plays the role of the relaxation time. When τ_{RC} *is* much shorter than the period of the square wave, the system reacts by recovering instantaneously to equilibrium, as typically observed in a merely viscous response, and $A(t)$ closely follows the slow perturbation $V(t)$.

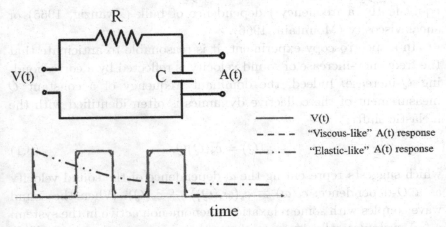

Figure 8-3:   A schematic representation of the analogy between a viscoelastic response and the one of RC circuit (see text).

Conversely, if $V(t)$ has period much shorter than τ_{RC}, the response of the system becomes too slow to equilibrate within the time-lapse of a single $V(t)$ oscillation. Clearly, this out-of-equilibrium response closely resembles the one of a fluid in the elastic limit.

When the square wave has a period comparable with τ_{RC}, it stimulates an intermediate response of the circuit, which depends on the actual period of $V(t)$ oscillations. This behavior resembles the intermediate "viscoelastic" response of the fluid.

8.2. Investigating viscoelastic phenomena by mesoscopic spectroscopy: The Q- and T-dependence of transport parameters

Coming back to spectroscopy experiments, viscoelastic effects on $S(Q,\omega)$ are best observable when the dominant frequency of density fluctuations becomes comparable with the inverse of the relaxation time. When this happens, the transport coefficients of the system become ω-dependent. More specifically, the viscoelastic transition is usually accompanied by a gradual decrease, upon ω-increase, of acoustic damping and a corresponding increase in sound speed.

Consistently, many theoretical descriptions of viscoelastic effects on the spectral shape relied on the inclusion of a time- (or, equivalently, a frequency-) dependence of bulk (Zwanzig, 1965) or shear viscosity (Mountain, 1966).

In a spectroscopy experiment, it is reasonable to anticipate that the frequency-increase of sound velocity is reflected by a corresponding Q-increase. Indeed, the dominant frequency of a constant Q measurement of the collective dynamics is often identified with the inelastic shift:

$$\omega_s(Q) = c_s(Q)Q, \tag{8.1}$$

which suggests representing the ω-dependence of the sound velocity as a Q-dependence: $c_s(\omega) \equiv c_s(\omega_s(Q)) \rightarrow c_s(Q)$. When the sound wave couples with some relaxation phenomenon active in the system, $c_s(\omega)$ systematically increases from the viscous to the elastic limit and the same increase is to be expected for the so-called apparent sound velocity $c_s(Q)$.[1] The Q- or ω-increase of the sound velocity is often referred to as the positive sound dispersion (*PSD*). In a typical spectroscopic measurement, this trend is revealed by a more-than-linear Q-increase of the inelastic shift, $\omega_s(Q) = c_s(Q)Q$.

More specifically, from low to intermediate Q, such a Q dependence is characterized by three regimes:

- In the viscous or hydrodynamic range, $\omega_s(Q)$ grows linearly with Q, as $\lim_{Q \to 0} \omega_s(Q) = c_0 Q$ (see Chapter 6), with c_0 equal to the adiabatic sound velocity c_s.
- Upon approaching the centre of the viscoelastic window, $\omega_s(Q)$ bends upwards following an "S"-shaped Q-dependence characterized by an inflexion point at $Q = Q^*$, where $\omega_s(Q^*) \approx 1/\tau(Q^*)$.
- Beyond this inflection point, *i.e.* in the elastic region, a new linear trend is retrieved, and its slope is determined by the elastic value of the sound velocity, c_∞.

In the simple scenario depicted above we neglected the coincomitant bending downwards of the dispersion curve stemming from the

---

[1]According to Eq. (8.1), this is true until until $c_s(Q)$ starts decreasing more rapidly than Q^{-1} which roughly happens for $Q > Q_m/2$, where Q_m is the position of the first sharp maximum of $S(Q)$.

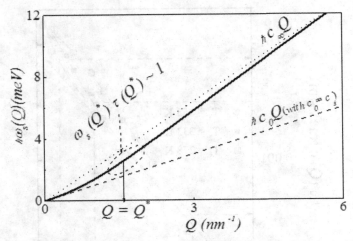

Figure 8-4: Schematic representation of the effect of a relaxation process on the dispersion curve. Throughout the text, this peculiar dispersive trend is commonly referred to as the positive sound dispersion, or PSD (see text). In the plot, the continuous line is the actual (also referred to as apparent) sound dispersion curve of the acoustic mode, its low and high Q linear trend being being indicated by the dashed and dotted respectively (see text). The dashed oval roughly locates the center of the viscoelastic crossover.

interaction of the acoustic waves with the local structure. Often, these bending effects become dominant at somewhat larger Q's, namely at values approaching $Q_m/2$, where Q_m is the position of the first sharp $S(Q)$ maximum. Given this *caveat*, the viscoelastic crossover of the dispersion curve is schematically represented in Figure 8-4.

An equivalent explanation of this dispersive trend can be gained from recognizing that the mere existence of a Q-crossover centred at $Q = Q^*$, lends itself to the introduction of a characteristic distance $\lambda^* = 2\pi/Q^*$. The latter can be identified with the size of the degree of freedom involved in the energy rearrangement triggering the relaxation process. When the response of the sample is averaged over distances either much larger or much smaller than λ^*, it resembles the one of a liquid or an elastic medium, respectively.

Let us now give a closer look at how the *PSD* manifests itself in a real experiment. The case of water seems particularly interesting, not much for the relatively high amplitude of *PSD* in

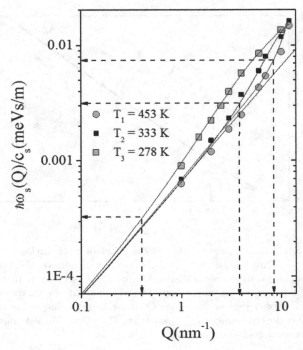

Figure 8-5:   Double logarithmic representation of the dispersion curves of water as determined from the inelastic shift of *IXS* spectra measured at the T's shown in the legend. The curves, taken from Ref. (Cunsolo, 1999, Cunsolo *et al.*, 1999) are divided for comparison by the adiabatic sound velocity derived from the Equation of State (Kestin *et al.*, 1984). Horizontal and vertical dashed arrows respectively locate inelastic shifts and exchanged wavevectors defining the center of the viscoelastic crossover (see text).

this system — a feature shared with noble gases — but rather for its marked temperature dependence.

This is clearly illustrated in Figure 8-5, which compares the sound dispersion curves derived from the inelastic shift of *IXS* spectra of water at three temperatures (278 K, 333 K and 453 K), as reported in (Cunsolo, 1999, Cunsolo *et al.*, 1999). The double logarithmic representation is here used to fully capture the viscoelastic crossover across the various orders of magnitude variation of the curves. Also, to make the comparison more meaningful, the dispersion curves are re-scaled by the corresponding adiabatic sound velocity, c_s, derived from the Equation of State (Kestin *et al.*, 1984). Notice that, with the performed normalization, if no *PSD* effect was present at all, all curves should have merged into the same linear trend, marked by the solid straight line. Conversely, the curves deviate from this simple linear behavior bearing somewhat lose evidence for inflexion points. The presence of a clear slope change can be perhaps better appreciated from the comparison with the corresponding polynomial best fits superimposed to the data as solid lines. These inflexion points, roughly located by vertical arrows in the plot, seem to shift to higher Q and $\omega_s(Q)$ values upon increasing T. This trend unveils the corresponding decrease of the relaxation time, *i.e.* a speed up of relaxation processes, which is a hallmark of structural relaxations, as also discussed in more detail further below.

8.3. Structural relaxations

Structural relaxations are probably the relaxation phenomena most thoroughly investigated by low-frequency spectroscopic techniques and are dominating in glass formers and associated liquids. In particular, the case of glycerol, a prototype of intermediate glass former according to the Angell classification scheme (Angell, 1985), has been in the focus of intensive experimental scrutiny, well before the advent of *IXS*. Amongst earliest investigations, the effects of a structural relaxation on density fluctuations of glycerol have been investigated by both Ultrasound Spectroscopy (*US*) (Piccirelli and Litovitz, 1957, Litovitz and McDuffie, 1963) and Brillouin Light Scattering (*BLS*) (Pinnow *et al.*, 1968, Montrose and Litovitz, 1970) techniques. These pioneering studies reported that structural relaxations in glycerol are characterized by a distribution of timescales becoming broader upon temperature decrease, which was initially ascribed to the onset of "cooperative jump processes". As the temperature of the liquid is lowered, the extent of intermolecular bonding increases and the free volume decreases, both effects contributing to slowing down internal rearrangements and enhance their cooperative character. Consequently, the dominant relaxation timescale, τ, sharply increases upon lowering the temperature, changing from $\approx 10^{-12}\,s$ in the regular, non-supercooled, liquid phase up to about 100 s close to the glass transition temperature.[2]

Being collective rearrangements of molecules hampered by the viscous drag, it can be expected that, at least in a first approximation, the timescale of these processes depends on the thermodynamic conditions as the viscosity does. In particular, as discussed in the next section, simple physical considerations lead to predicting that the structural relaxation time steeply increases with the inverse of the temperature, following an exponential or an even sharper growth. Structural relaxations are sometimes referred to as α-relaxations, and their relevant phenomenology has been formally described in Refs. (Gotze, 1991, Gotze and Sjogren, 1995).

An example of the interplay between temperature and frequency dependencies unveiling the presence of a structural relaxation is provided by Figure 8-6; the plot displays the frequency-dependence of the sound velocity of glycerol measured at various temperatures by complementary methods. Experimental values are taken from a joint experiment (Giugni and Cunsolo, 2006) combining *IXS* measurements, ranging in the terahertz window, with *BLS* and Brillouin Ultra Violet Scattering (*BUVS*) ones, spanning the 1.5–20 GHz range. Furthermore, these results are compared with previous literature *US* data spanning the 10–150 MHz range (Jeong, 1987, Jeong *et al.*, 1986). Following a customary representation, the various spectroscopy results are plotted as a function of the frequency $\nu = \omega/2\pi$, where the angular frequency ω is identified with the dominant acoustic frequency determined by each measurement, i.e. it is assumed $\omega = \omega_s(Q)$.

As expected, the sound velocity sharply increases upon increasing the frequency of the spectroscopic probe, while the inflexion points of the curves shift to higher frequencies upon temperature increase.

As an alternative pathway to characterize viscoelastic phenomena, one can look instead at the temperature dependence of the

---

[2]The glass-transition temperature T_g is somehow arbitrarily defined, as it depends on the cooling rate of the glass-forming system. In the attempt to rationalize such arbitrariness, the value usually considered for T_g is temperature at which the system, at a conventional cooling rate of about 10 K/min, attains a relaxation time of $\approx$100 s and a viscosity 10^{12} Pa $\cdot$ s.

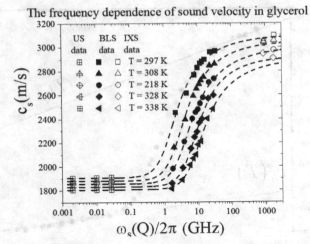

Figure 8-6: The sound velocity of glycerol measured at different frequencies by complementary techniques and various temperatures, as shown in the legend. The dashed lines are best fits to data, as obtained by assuming a continuous distribution of relaxation time scales (Giugni and Cunsolo 2006). Data are redrawn from Ref. (Giugni and Cunsolo, 2006), which also includes *US* measurements from Refs. (Jeong *et al.*, 1986) and (Jeong, 1987).

acoustic mode parameters while keeping fixed Q or, more in general, the frequency of the spectroscopic probe. To give a practical example, Figure 8-7 illustrates the sharp temperature dependence of the sound velocity of glycerol at its viscoelastic crossover. These low-frequency spectroscopic results are taken from *US* measurements (Jeong, 1987, Jeong *et al.*, 1986), a part of which was previously included in Figure 8-6.

It clearly appears that reported data cover the whole transition of the sound velocity between the low-T elastic limit and the high-T viscous one, these two limiting behaviors being reported as solid lines in the plot for reference.

In most cases, the pure viscoelastic or single timescale hypothesis is an oversimplification, and assumptions based on the coexistence of multiple timescales or even a continuous distributions of them (Williams and Watts, 1970, Phillips, 1996, Davidson and Cole, 1951) are more realistic. As a matter of fact, in Ref. (Giugni and Cunsolo, 2006) it was found that the frequency-dependence of sound

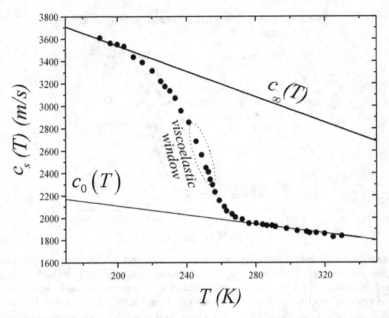

Figure 8-7:   The temperature dependence of the sound velocity of glycerol (dots) as measured by *US* techniques in Ref. (Jeong, Nagel *et al.*, 1986). The straight lines representing $c_0(T)$ and $c_\infty(T)$ are reported as determined through the high and low T slopes of the $c_s(T)$ curve.

velocity of glycerol is described more accurately by a Cole-Davidson model (Davidson and Cole, 1951) — represented by dashed lines in Figure 8-6 — rather than a simple viscoelastic trend.

8.4. Brief remarks on the temperature dependence of relaxation time

As mentioned in the previous section, a distinctive feature of structural relaxation processes is the sharp temperature dependence of their timescale, which roughly parallels the one of viscosity. A general idea of the temperature dependence of the viscosity stems from straightforward considerations. As a first approximation, one can assume that the viscosity is directly related to the inverse of the diffusion coefficient through the Stokes-Einstein formula (Einstein 1956) and also assume that diffusive processes are activated by single

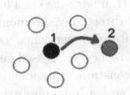 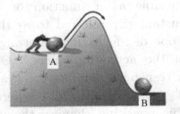

Figure 8-8: Conceptual representation of an activated process: the jump diffusion and a conceptual macroscopic analogy (see text).

atomic site jumps. As illustrated in the pictorial scheme of Figure 8-8, to jump from the site "1" to the site "2", an atom must overcome a potential barrier primarily determined by the cage of first neighbours. Macroscopically, this situation suggests a vague analogy with the case of a man pushing a stone in the attempt to move it from A to B (left part of the figure). The accomplishment of such a movement requires that the work made by the man exceeds the potential barrier originating from the hill ahead (asides of the expectedly huge friction resistance!).

Similarly, the probability of the atomic jump between the sites 1 and 2 proportional to $\exp(-W/k_B T)$ where W is the energy threshold to be overcome to activate the jump (Eyring, 1936). Assuming that the diffusion entirely consists of these out-of-cage elementary jumps, it can be deduced that the diffusion coefficient D is proportional to such a probability. According to the Stokes-Einstein formula, the (shear) viscosity $\eta_s \propto D^{-1}$; therefore, in a first approximation one has $\eta_s = \eta_0 \exp(W/(k_B T))$. Assuming that relaxation time and viscosity have the same activation threshold, one can express their T-dependence in the following general form:

$$k = k_0 \exp\left(\pm W/k_B T\right), \tag{8.2}$$

where the parameter k can be identified either with the viscosity or the relaxation time. By taking the logarithm of both members of the above equation, it can be readily concluded that, for activated processes, the logarithm of the viscosity (or the relaxation time) depends linearly on $1/T$. The semi-logarithmic plot of a given variable as a function of $1/T$ (or, more often, of $1000/T$) is customarily referred to as the Arrhenius plot (Laidler, 1987). The slope of the Arrhenius straight line provides a direct measurement of the activation energy. Often, the temperature dependence of a transport parameter, or a reaction rate, is rather expressed as $k = k_0 \exp(\pm W/RT)$ where R is the gas constant. Obviously, this rewriting changes the unit in which the activation energy is expressed; this must be the same as for RT, *i.e.* typically kcal/mole.

In many fluids, especially the highly viscous ones, the dependence of $\log k$ on $1/T$ is sharper than Arrhenius. In a first approximation, this circumstance can be explained by recognizing that a diffusive jump can only happen if there is a void space large enough to accommodate it, so the jump probability should depend on the free-volume.

To set the basis for a more formal treatment (Cohen and Turnbull, 1959), one can assume that the free volume per molecule has a continuous distribution. Such a free volume is defined as the volume enclosed by first neighbors cage minus the one of the molecule itself v_0. One can also define an average free volume as $v_f = V_f/N$, with V_f being the total free volume of the N molecules' system. In this theory the diffusion coefficient as $D = D'(v_0)P(v_0)$, where $P(v_0)$ is the probability that the molecule finds a free volume at least as large as v_0. Cohen and Turnbull found that this probability is $P(v_0) = \exp(-\Gamma V_0/v_f)$, where Γ is a numerical factor introduced to correct for the molecular free volume overlaps and is included between 0.5 and 1 while the volume V_0 approaches the close-packed molecular volume. Similar free volume theories demonstrated a remarkable accuracy in describing the pressure dependence of experimental data, although they failed to predict their temperature dependence.

The complementary character of the Eyring and Cohen and Turnbull theories prompted the idea of Macedo and Litovitz (Macedo and Litovitz, 1965) of combining them. Within this approach, the Arrhenius formula is modified to account for the dependence of diffusive jumps on both activation energy and free volume available.

Correspondingly, the viscosity is written as:

$$\eta_s = \eta_0 \exp\left(W/k_B T\right) \exp\left[\Gamma V_0/(V - V_0)\right], \qquad (8.3)$$

where we considered that the total free volume is $V_f = N v_f = V - V_0$, with $V_0 = N v_0$. The formula above demonstrated to provide a reasonably consistent description of viscosity measurements in a large variety of liquids, across a broad range of thermodynamic conditions.

Within the model of Machedo and Litovitz V_0 does not depend on T, *i.e.* volume variations with T uniquely involve the free volume. Given that the expansion coefficient $\alpha(T) = \partial\left[ln(V)\right]/\partial T$ one can write:

$$V - V_0 = V_0 \left[\int_0^T dT \alpha(T) - 1\right] \approx \alpha V_0 (T - T_0), \qquad (8.4)$$

where the last equality stem from the simplified hypothesis that the expansion coefficient does not depend on T, *i.e.* $\alpha(T) \equiv \alpha$.

This simple description leads to conclude that, for systems in which $W/(k_B T) \ll V_0/V_f$ temperature dependence of the viscosity or the relaxation time has the following form:

$$k = k_0 \exp\left[A/(T - T_0)\right]. \qquad (8.5)$$

Where the constant $A = \pm \Gamma V_f/(\alpha V_0)$ has the dimension of a temperature. The temperature dependence defined above is substantially sharper than predicted by the Arrhenius law, due to the circumstance that the exponent diverges at a finite temperature T_0. A temperature trend having the form of the above equation is customarily referred to as Vogel Tamman Fulcher (Fulcher, 1925) law. For instance, it describes accurately the temperature evolution of the so called fragile glass formers (Angell, 1995). A similar result can also be obtained from a general thermodynamic treatment of a system in which cooperatively arranging region coexist (see Debenedetti, 1996).

8.5. Quantitative insight on the structural relaxations: The case of water

Across the last five decades, spectroscopic methods covering different dynamic domains have been thoroughly employed to investigate viscoelastic properties of water, in both liquid and supercooled phases. These include:

(i) *US* measurements covering the 10 KHz-100 MHz frequency window and wave-vector roughly ranging in the $10^{-8} - 10^{-4}$ nm^{-1} interval (Slie, 1966, Davis and Jarzynsky, 1972, Trinh and Apfel, 1980);

(ii) *BLS* measurements covering the 1–10 GHz frequency window with exchanged Q's spanning the 10^{-2} nm^{-1} range (Rouch, 1976, Rouch *et al.*, 1977, Magazu *et al.*, 1989, Cunsolo and Nardone, 1996) and, more recently, Brillouin Ultra Violet Scattering (BUVS) (Bencivenga *et al.*, 2009) and time-resolved light techniques (Torre *et al.*, 2004).

Despite the largely different dynamic windows and the non trivial experimental challenges, most of these spectroscopy investigations provided a fairly consistent picture of the viscoelastic behaviour of water dynamics.

Main experimental challenges primarily stemmed from the difficult observation of viscoelastic effects in a mildly viscous fluid as liquid water. In this system, internal rearrangements are too fast to be efficiently probed by ultrasonic (kilohertz to megahertz) and hypersonic (gigahertz) techniques unless the sample is deeply supercooled. Nonetheless, even at the lowest T's reachable with supercooled bulk water samples (say around 247 K), the relaxation time τ can at most reach the 10 *ps* window, thus being still too short to fall within the domain of sensitivity of *BLS*, or, even less, the one of *US*.

For this reason, in the various low-frequency measurements on water reported in the literature, the quantitative determination of the structural relaxation time required either bold ω-extrapolations (Magazu *et al.*, 1989), or the mixing with a more viscous fluid. The latter is the case of an *US* absorption study in the mid-1960s (Slie, 1966), in which relaxation timescales in water were derived by extrapolating the results obtained in increasingly diluted aqueous solutions of glycerol.

The case of water is also particularly interesting for the scope of this book as a host of independent *IXS* measurements demonstrated that the relaxation process coupling with the spectrum of density fluctuation of water preserves its structural character down to mesoscopic (nm, ps) scales (Cunsolo *et al.*, 1999, Monaco *et al.*, 1999, Pontecorvo *et al.*, 2005, Bencivenga *et al.*, 2007a, Bencivenga *et al.*, 2007c).

8.5.1. *Gaining insight on the relaxation process from the spectral shape*

In principle, an estimate of the relaxation timescale can be achieved by looking at the Q-dependence with the reduced longitudinal modulus:

$$M_r(Q) = \frac{\omega_s^2(Q) - \omega_0^2(Q)}{\omega_\infty^2(Q) - \omega_0^2(Q)}$$

which varies from zero to one as the dispersion curve $\omega_s(Q)$ evolves from its viscous and elastic limits, *i.e.* $\omega_0(Q)$ and $\omega_\infty(Q)$, respectively. At the center of the viscoelastic crossover $Q^\star$ one has $\omega_s(Q^\star) = 1/\tau(Q^\star)$ and $M_r(Q^\star) = 0.5$. The $M_r(Q^\star) = 0.5$ condition can be located by linear interpolation of $M_r(Q)$; the successive interpolation of $\omega_s(Q)$ in the point $Q^\star$ enables a direct measurement of the relaxation time through $\tau(Q^\star) = 1/\omega_s(Q^\star)$.

A similar strategy was followed in a *IXS* measurement on water (Cunsolo *et al.*, 1999), dealing with the dependencies of the reduced longitudinal modulus on both T and Q. This method led to draw some important conclusions on terahertz relaxation phenomena in water demonstrating both its structural character and its link with the process of breaking and forming of hydrogen bonds. However the success of this method ultimately rested on suitable assumptions on on the values of $\omega_0(Q)$ and $\omega_\infty(Q)$ and the use of a lineshape model, a *DHO* profile integrated with a lorentzian central peak, providing a rigorous account of the spectral shape in the viscous and elastic limits only.

More in general, the measurement of the sound dispersion curve $\omega_s(Q)$ might be model-dependent or being of limited accuracy around the centre of the viscoelastic crossover, $Q \approx Q^*$, where inelastic and quasielastic feature might become highly damped and loosely resolved.

Gaining direct insight on the relaxation time and other relaxation-related quantities should rather be directly pursued by best fitting the measured spectra with a model accounting explicitly for viscoelastic effects on the spectral profile. As extensively discussed in Chapter 6, the memory function formalism enables to build up directly such a lineshape model. Most memory-based models used to account for viscoelastic effects on *IXS* spectra are obtained as suitable approximations of the following Molecular Hydrodynamics (*MH*) second-order memory function:

$$m_L(Q,t) = \nu_L(Q,t)Q^2 + \omega_0{}^2(Q)(\gamma(Q)-1)\exp(-\gamma(Q)D_T(Q)Q^2 t)$$

$$(8.6)$$

where:

$$\nu_L(Q,t)Q^2 = 2\Gamma_0(Q)\delta(\tau) + \Delta^2(Q)\exp[-(c_0^2(Q)/c_\infty^2(Q))t/\tau_C(Q)],$$

$$(8.7)$$

with $\Gamma_0(Q)$ and $\Delta^2(Q) = Q^2[c_\infty^2(Q) - c_0^2(Q)]$ being the strengths of the Markovian and the relaxation terms, respectively.

In Eq. (8.7) we have introduced the generalized compliance relaxation timescale, defined as $\tau_C(Q) = [c_\infty^2(Q)/c_0^2(Q)]\tau(Q)$. When modeling *IXS* spectra, the use of $\tau_C(Q)$ is often preferred to the one of $\tau(Q)$ as it enables a straightforward comparison with the outcome of low-frequency spectroscopic measurements, which can directly determine $\tau_C(Q = 0)$ as the inverse of the halfwidth of the so-called Mountain mode (Mountain, 1966), which is a quasielastic spectral feature fingerprinting the coupling of $S(Q,\omega)$ with an active relaxation. Furthermore, the use of the compliance relaxation time makes *IXS* results directly comparable with the measurements of other susceptibility spectra such as, *e.g.*, depolarized light scattering, dielectric relaxation, and so forth. From a physical point of view, $\tau_C(Q = 0)$ represents the relaxation time of the compliance J, which is linked to the longitudinal modulus as $J = M^{-1}$ (Herzfeld and Litovitz, 1965).

In practice the number of free parameters in the model defined by Eqs. (8.6) and (8.7) can be reduced by suitable approximations pertinent to the specific dynamic (Q, ω)-range or to the thermodynamic conditions of the sample.

Also, the superimposition of the sum rules fulfilment helps to further reduce the number of adjustable parameters in the lineshape analysis, besides improving the physical soundness of the model.

8.5.2. *An IXS measurement of the structural relaxation time of water*

Let us now give a closer look at the results of an *IXS* work on water covering a broad thermodynamic region with an almost constant density path (Monaco *et al.*, 1999). In such a work, the structural relaxation time of water was derived from a modeling of *IXS* spectra based upon an *MH* model of the memory function (see Eqs. 8.6–8.7), in which all parameters of the Rayleigh thermal contribution were kept fixed to the values derived from the Equation of State (Kestin *et al.*, 1984).

The relaxation timescales of water derived from this lineshape modeling are reported in Figure 8-9 and they show that, consistent with expectations, $\tau_C(Q)$ decreases substantially either upon T-decrease or Q-increase.

It is important to recognize that, besides the discussed temperature dependence, one of the hallmarks of structural relaxation phenomena is the pronouncedly Q-dependent timescale, which stems from the cooperative nature of these processes. Indeed, the response of the fluid probed by a spectroscopic experiment is averaged over a volume $\sim Q^{-3}$, which becomes gradually smaller upon increasing Q. Therefore, an increase of Q ultimately enables one to explore structural rearrangements involving a decreasing number of molecules which can be thus be accomplished in shorter lapses.

In Ref. (Monaco, Cunsolo *et al.*, 1999) a linear model was used to approximate the Q-decay of the relaxation time, and the corresponding linear best fits are also included in Figure 8-9.

In principle, a linear assumption for the Q-dependence lacks rigorous ground. In fact, the Q-derivative of transport parameters

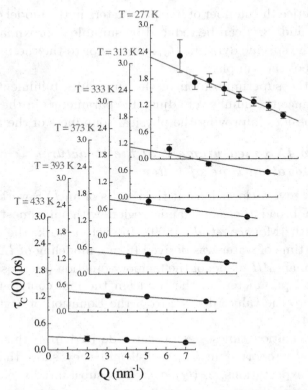

Figure 8-9:   The Q-dependence of the relaxation timescale $\tau_C(Q)$ of water is reported as derived from Ref. (Monaco, Cunsolo *et al.*, 1999) at the temperatures indicated in the individual plots. The corresponding linear best fits used to extrapolate the $Q = 0$ limits are also reported as solid lines for reference.

must vanish upon approaching the hydrodynamic regime, as therein these parameters should become Q-independent, or global, properties of the system. In this respect, a parabolic, Lorentzian, or Gaussian Q-dependence seem more appropriate heuristic assumptions. However, if the Q range is sufficiently small, a linear law is a reasonably accurate approximation entailing a limited number of adjustable parameters. In some cases (Bencivenga *et al.*, 2007a, Bencivenga *et al.*, 2007b), an exponential decay was assumed instead.

One of the primary purposes for fitting the Q-dependence of relaxation timescale $\tau_C(Q)$ is the opportunity to extrapolate the

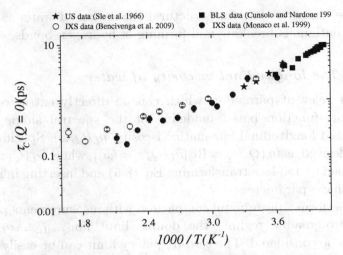

Figure 8-10: The Arrhenius plot of the $Q = 0$-extrapolated relaxation time of water as measured by various techniques as indicated in the legend.

macroscopic $(Q = 0)$ values of this parameter, which coincide with the intercepts of the best fitting straight lines in Figure 8-9. These extrapolations are reported in Figure 8-10 in an Arrhenius plot and therein compared with:

(1) the relaxation times extrapolated from *IXS* measurements on water spanning both subcritical and supercritical temperatures (Bencivenga *et al.*, 2007a, Bencivenga *et al.*, 2007b),
(2) those extrapolated from *US* measurements on increasingly diluted glycerol-water mixtures (Slie, 1966), and, finally:
(3) those measured by *BLS* in the supercooled phase (Cunsolo and Nardone, 1996).

Overall, experimental results in Figure 8-10 show that the relaxation time of water follows a linear Arrhenius trend, at least in the 250 K to 500 K *T*-range covered by experimental data. In particular, the the linear fit of the Arrhenius plot of *IXS* data in Ref. (Monaco, Cunsolo *et al.*, 1999) yields an activation energy of 3.8 kcal/mol, in agreement with the known value of the hydrogen bond energy estimated through Raman measurements (Walrafen, 1972). This evidence suggests that the structural relaxation of water is linked to the continuous breaking and forming of hydrogen bonds.

8.5.3. *The longitudinal viscosity of water*

Another relevant parameter which can be directly extracted from a memory-function based modeling of the spectral shape is the generalized longitudinal kinematic viscosity, $\nu_L(Q,\omega)$. Specifically it can be derived as $\nu_L(Q,\omega) = \mathrm{Re}\,[\tilde{\nu}_L(Q, z = i\omega)]$, where $\tilde{\nu}_L(Q, z = i\omega)$ is obtained by Laplace transforming Eq. (8.5) and inserting in it best-fitting shape parameters.

To perform a meaningful comparison with measurements probing the hydrodynamic regime, the double limit $\lim_{Q\to 0,\omega\to 0}\nu_L(Q,\omega)$ needs to be considered. The zero-frequency limit can be easily evaluated analytically and yields the following Q-dependent generalization of the longitudinal viscosity:

$$\nu_L(Q) = \left[2\Gamma_0(Q) + \Delta^2(Q)\frac{c_0^2(Q)}{c_\infty^2(Q)}\tau_C(Q)\right]. \qquad (8.8)$$

As previously discussed for the relaxation time, the $Q = 0$ value of $\nu_L(Q)$ can be determined by extrapolation by best fitting its Q-dependence. In Figure 8-11 the $Q = 0$ extrapolations derived from *IXS* results on water (Monaco *et al.*, 1999, Bencivenga *et al.*, 2007a, Bencivenga *et al.*, 2007b) are reported and therein compared with the macroscopic value of the shear component of the kinematic viscosity ν_s, as derived from the Equation of State in Ref. (Kestin, Sengers *et al.* 1984) after rescaling, for the sake of comparison, by the ν_L/ν_s ratio deduced from *US* measurements in the literature (Davis and Jarzynsky, 1972).

From the plot, it readily appears that at sufficiently high temperatures the $Q = 0$ value of the viscosity $\nu_L(Q = 0)$ follows a nearly linear Arrhenius trend at least at high temperature. This trend leads to estimate an activation energy comparable to the one derived for the relaxation timescale and with the hydrogen bond lifetime.

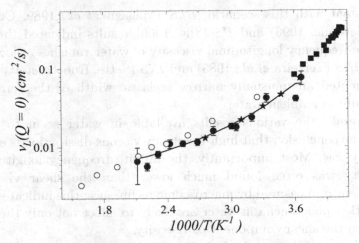

Figure 8-11: The Arrhenius plot of the kinematic longitudinal viscosity of water as determined by different spectroscopic techniques, with the same symbols as in Figure 8-10. The macroscopic value deduced by combining *US* measurements (Davis and Jarzynsky, 1972) and the Equation of State in Ref. (Kestin, Sengers *et al.*, 1984) is reported as a solid line for reference.

Notice that at lower T values the curve might be compatible with an increase sharper than a simple Arrhenius trend; however, it seems still an open question whether this trend, also documented by other lower frequency measurements, has the form of a Vogel-Tamman-Fulcher law or the one of a critical-like divergence. The latter hypothesis seems consistent with an Optical Kerr Effect (*OKE*) measurement reporting a critical like divergence of the relaxation time (Torre *et al.*, 2004).

8.5.4. *The microscopic contribution to the viscosity*

The viscoelastic analysis of *IXS* spectra of water in Ref. (Monaco, Cunsolo *et al.*, 1999) also sheds insight on the amplitude of the instantaneous term, $2\Gamma_0(Q)$ in Eq. (8.7) which, as discussed, accounts for the fast microscopic contribution to the time decay of the memory function. Although relatively small at moderate T's, the amplitude of this term substantially increases with increasing T, changing from 5% of the total viscosity at 273 K up to nearly 40% at 473 K.

Consistent with this scenario, *BLS* (Magazu *et al.*, 1989, Cunsolo and Nardone, 1996) and *US* (Slie, 1966) results indicated that the infinite-frequency longitudinal viscosity of water vanishes at lower T, while *INS* (Teixeira *et al.*, 1985) and *IXS* (Sette, Ruocco *et al.*, 1995) documented an unusually narrow inelastic width of the terahertz spectrum of ambient water.

Overall, the various results available in water seem to agree upon the conclusion that high-frequency viscous dissipation is exceptionally low. Most importantly, the infinite frequency longitudinal viscosity was often found much lower than the shear viscosity coefficient η_s measured by macroscopic techniques; this indicates that relaxation phenomena in water are likely to affect not only the bulk but also the shear component of viscosity.

It has to be recognized that an almost entirely relaxing shear viscosity seems consistent with the ability of water to support shear propagation at moderate Q values, and the emergence in its $S(Q,\omega)$ of a transverse acoustic mode (Cunsolo, 2015, Cunsolo *et al.*, 2016).

Concerning the strength of the microscopic contribution, $\Gamma_0(Q)$, a qualitative analysis of data reported in Ref. (Monaco *et al.*, 1999), suggests that its monotonic Q-increase is well-approximated by a quadratic law:

$$\Gamma_0(Q) = D_0 Q^2.$$

Indeed, all *IXS* results on water (Bencivenga *et al.*, 2007a, Bencivenga *et al.*, 2007b, Monaco *et al.*, 1999) essentially confirmed what is usually reported for glass-forming systems, *i.e.* that this fast contribution increases as Q^2. Notice that in glass-forming materials, $\Gamma_0(Q)$ primarily determines the linewidth of the *Brillouin* line, which, indeed has been observed to grow as the square of Q (Sette *et al.*, 1998). Parabolic growths best fitting the Q-dependence of $\Gamma_0(Q)$ provide a direct determination of D_0, whose value, in water, is in the 10^{-3} cm^2/s range and seems reasonably insensitive to thermodynamic changes (Bencivenga *et al.*, 2007a, Bencivenga *et al.*, 2007b), in accord with the terahertz damping measured in glass-forming systems near the melting or in supercooling states (Sette *et al.*, 1998). Concerning this last point, it is useful to recognize that, in these highly viscous systems, the condition $\omega_s(Q)\tau(Q) \gg 1$ in the dynamic range probed by *IXS*, where the damping is mostly determined by the high-frequency microscopic dynamics.

8.6. Collisional relaxations

After discussing structural relaxations with a special focus on the case of water, here we deal with collisional relaxations, which relate to the rapid rattling of single molecules inside the transient cage formed by their first neighbours. It is customarily assumed that oscillations of single molecules around their "lattice" sites remain confined within the first neighbouring cage until the latter rearranges due to the structural relaxation. In Chapter 2 (see Section 2.2) we also showed that the period of *in-cage* molecular oscillations is expectedly comparable to the inverse of the Einstein frequency (Eq. (2.21)) which typically span the sub-picosecond range.

In summary, within the relaxing cage scenario, one can assume that short-time relaxations directly relate to these rapid oscillatory movements induced by successive collisions between the tagged molecule and its first shell neighbors. At variance with structural relaxations, these collisional relaxation processes have a mildly temperature-dependent timescale, as their non-cooperative nature makes them less prone to the viscous drag.

A direct assessment of the collisional character of a relaxation process could arise from a comparison of the related timescale with the particle intercollision time. As an estimate of the latter timescale, one could, in principle, take the inverse of the Einstein frequency; unfortunately, its explicit evaluation is not very straightforward, as it would require the knowledge of the interatomic potential and the pair correlation function (see Eq. (2.21)). A simpler reference parameter is the hard-sphere inter-collisions time, τ_{coll}. Within a hard-sphere model, the inter-collision time depends on the mass and size of the atom as well as on the temperature of the system, through the simple formula:

$$\tau_{coll} = 1/n\sqrt{M/\left(16\pi\sigma_{HS}^4 k_B T\right)}, \tag{8.9}$$

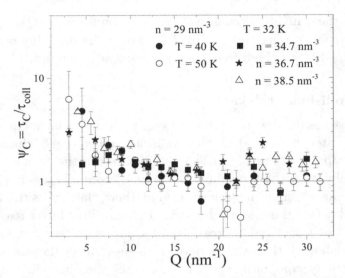

Figure 8-12: The Q-dependence of the reduced relaxation time $\psi_C(Q) = \tau_C(Q)/\tau_{\text{coll}}$ (see text), as obtained from *IXS* measurements on Ne at the thermodynamic conditions indicated in the plot. Data are obtained from results reported in Ref. (Cunsolo *et al.*, 2001). The horizontal line marks the constant limiting value characteristic of the collisional regime.

where σ_{HS} is the hard-sphere diameter. A simple derivation of this formula is proposed in Appendix 8A.

To assess a possible link between the relaxation time derived from the *IXS* spectra of a simple monatomic fluid and the inter-collision time defined by Eq. (8.9), one can refer to Figure 8-12. There, the Q-dependence of the ratio $\psi_C(Q) = \tau_C(Q)/\tau_{coll}$ in neon is reported in semi-logarithmic plot, as evaluated using the $\tau_C(Q)$ values derived from the viscoelastic modeling of *IXS* spectra discussed in Ref. (Cunsolo *et al.*, 2001).

Despite some scattering of data, it appears that for Q larger than about 8 nm^{-1}, $\psi_C(Q)$ approaches a nearly constant value somewhat close to one, as to be expected for a merely collisional relaxation. Collisional relaxations, at variance with their structural counterparts, are expectedly characterized by a timescale mildly changing with Q as a reflection of their non-collective nature, i.e. their substantial invariance upon changes of the probed volume ($\propto Q^{-3}$). Conversely, the onset of a Q-dependence can be appreciated in the low-Q values of ψ_C in Figure 8-12. This trend might reveal an increasing relevance of structural relaxations upon approaching macroscopic distances, *i.e.* the $Q = 0$ limit.

From the plot it appears that the relaxation loses any relic of a Q-dependence at about half the position of first sharp diffraction peak, which is at about $23\,\mathrm{nm}^{-1}$ for neon.

8.7. Other types of relaxation phenomena

As perhaps evident from the discussion above, the physical origin and the phenomenology of a relaxation process depend in principle on the specific degrees of freedom involved in this phenomenon, as well as the nature of intermolecular interactions. For instance, if the molecules have pronounced asymmetry in their mass distribution, a relaxation may arise from the interplay between translational and rotational degrees of freedom of such molecules, customarily referred to as rotational-translational coupling, or *RTC*. From a theoretical point of view, the effects of the *RTC* on the spectral density can be described by inserting an additional tensor variable in the hydrodynamic description of the system's response, which accounts for the anisotropy of the intermolecular potential. Eventually, this leaves a signature on the spectral shape similar to the one stemming from the coupling with a relaxation process (Keyes and Kivelson, 1971).

Liquid Salol (phenyl salicylate) is probably the system in which the onset of an *RTC* has been most thoroughly scrutinized. This phenomenon was probed by both polarized (Keyes and Kivelson, 1971, Dreyfus *et al.*, 1992) and depolarized (Pratesi *et al.*, 2000, Cummins *et al.*, 1996) light scattering measurements as well as by time-resolved techniques, such as transient grating impulsive stimulated thermal scattering (Glorieux *et al.*, 2002), or Optical Kerr Effect Spectroscopy (Bohmer *et al.*, 1996, Torre, 2007). The latter methods implement the "pump-probe" scheme, in which a probed beam is used to generate a perturbation bringing the sample out of equilibrium, the successive recovery of the sample to equilibrium is probed by looking at the time-dependent scattering.

In most of these works the effect of the coupling between degrees of freedom has been described using the Mode-Coupling Theory (Gotze and Sjogren, 1992) of the liquid glass transition. In particular, many experiments were aimed at testing the presence of a dynamical transition, at some temperature T_0 above the actual temperature of the glass transition T_g.

As the relaxation phenomena are caused by energy redistributions between density fluctuations and internal movements of the system, their phenomenology can be, in principle, as diverse and complex as the microscopic dynamics. However, it is customary to group all relaxation phenomena into a limited number of broad categories, like those introduced in previous section (microscopic, structural, rotational and so on); indeed, it is often hard to unambiguously identify the specific degrees of freedom involved, due to both experimental difficulties and the limited discrimination ability of the modeling. These categories are often based on phenomenological aspects as the dependence on thermodynamic conditions and on the exchanged wavevector Q, which unveil either the cooperative or the microscopic character of the phenomenon.

Finally, it is broadly recognized that relaxation processes can also have a non-dynamic origin and arise from the intrinsic topological disorder of amorphous materials. To better understand this point, it is useful to refer to a more straightforward case, where the density waves propagate through a perfectly ordered sample, such as an ideal crystal.

The transfer of momentum $\hbar Q$ to a crystal moves it in a well-defined eigenstate of the momentum, the phonon, characterized by a plane wave with wavenumber Q and frequency ω_s. Conversely, when a momentum $\hbar Q$ is transferred to an amorphous material, due to the lack of structural order in the target system, the atoms oscillate in different directions, that is the microscopic eigenmodes have different projections on Q. Consequently, the eigenstates of the momentum in a disordered system can no longer be represented as plane waves (Ruocco *et al.*, 2000), their wavefronts being instead somewhat distorted surfaces. The interplay of these waves causes a mutual dephasing and ultimately shortens the lifetime of the dominating collective excitation.

In some respects, the effect of this dephasing is equivalent to a relaxation phenomenon affecting the acoustic wave, but at variance of an ordinary relaxation, this process has a merely "topological", rather than dynamical, origin, as it stems from the inherent structural disorder of amorphous systems. In glasses this effect yields the dominating contribution to sound attenuation (Ruocco *et al.*, 1999).

Concerning simpler systems, it is worth mentioning an *IXS* work on molten potassium (Monaco *et al.*, 2004) which reported the presence of a sizeable *PSD* and ascribed it to the inherent topological disorder of the liquid. This assignment rested on the assumption that decays channels active in potassium are uncoupled from structural rearrangements, typically "frozen" over the short timescale probed by *IXS*.

In general, this microscopic contribution to the viscous decay is accounted for by the instantaneous or Markovian term in the memory function, which, in principle, in molecular fluids also accounts for ultrafast intramolecular vibration inducing early-stage time decays of the memory function, usually spanning timescales shorter than 0.1 ps.

8.8. The adiabatic-to-isothermal transition

Let's now focus on a different coupling mechanism affecting density fluctuations and giving rise to a negative sound dispersion of the acoustic wave, i.e. a dispersive effect opposite to that typically associated with viscous relaxation phenomena.

As usual, we can resort to the description of the sound wave as a train of compression-rarefaction zones (*CRZ*s), which "throb" with the same period as the acoustic wave. Furthermore, to ease the notation, we will drop any reference to the Q-dependence of generic transport or relaxational parameters X, i.e. we will implicitly assume $X \equiv X(Q)$.

We can start from considering that the excess of energy stored in a compression zone is then dissipated toward the propagation background via the thermal diffusion mechanism. For small Q's, the timescale of such a thermal dissipation, $\tau_T = 1/(\gamma D_T Q^2)$, undergoes a Q^{-2}-divergence, which is steeper than the Q^{-1}-divergence of the acoustic period, $T_{AW} = 2\pi/(c_s Q)$. This implies that, for most fluids, low Q thermal exchanges are much slower than acoustic oscillations. Consequently, acoustic propagation takes place without thermal losses, i.e. adiabatically (**adiabatic** regime), and the CRZs keep an internal temperature different from the one of the propagation background.

Upon Q-increase, τ_T decreases more rapidly than T_{AW}, therefore the adiabatic approximation gradually breaks down, and thermal dissipation starts becoming sizable within time lapses comparable with the period of a single acoustic oscillation.

Consequently, for Q values further away from the adiabatic limit the situation might be gradually reversed, and a new **isothermal** regime can be joined in which thermal exchanges affecting acoustic propagation become much more rapid than acoustic oscillations. Here, it can be safely assumed that the CRZs of the acoustic waves are always in thermal equilibrium with the propagation environment.

Due to the increasing relevance of thermal dissipation, the sound velocity systematically decreases at the adiabatic-to-isothermal (AI) Q-crossover, thus exhibiting a negative sound dispersion.

A sensible parameter to identify the adiabatic or isothermal character of the acoustic propagation probed by a scattering experiment is the timescale ratio $\xi = T_{AW}/(2\pi\tau_T) = D_T Q/c_s$. In reference to this metric, adiabatic and isothermal limiting regimes are characterized by the conditions $\xi \ll 1$ and $\xi \gg 1$, respectively.

In this perspective, especially interesting are systems having high thermal diffusivity as liquid metals, for which the AI crossover can be observed at low or moderate Q values, where the acoustic mode disperses linearly. Unfortunately, in liquid metals, the observation of this crossover is somewhat challenging. In fact, across the AI transition, the sound velocity decreases from the adiabatic (c_s) to the isothermal value ($c_s/\sqrt{\gamma}$) and the relative amplitude of this decrease is rather small for liquid metals, which typically have $\gamma \approx 1$. Such an amplitude is instead sizable in noble gases — for which γ values larger than 2 are not uncommon; however, the substantial acoustic damping in these systems hampers a precise determination of the sound velocity, thus challenging the observation of an *AI* crossover.

From the experimental side, an *AI* transition was observed in non-conductive fluids well beyond continuous scales, i.e. in the mesoscopic regime probed by *IXS*, in which the Q-dependence of transport parameters can by no means be neglected. Also, the samples where in nearly critical or supercritrical condition where the sought for effect is best visible thanks to relatively large values of both D_T and γ.

Another key variable directly characterizing an *AI* crossover is the reduced modulus:

$$M_T = \frac{\omega_s^2 - \omega_T^2}{(\gamma - 1)\omega_T^2},$$

where ω_s and ω_T represent respectively the measured frequency of the acoustic mode and its isothermal value. Notice that the denominator of M_T coincides with the amplitude of the Rayleigh contribution to the memory function decay. Its value can thus be determined through a lineshape best-fitting with an *MH* of the memory function as the one defined in Eqs. (8.4–8.5). By definition, M_T varies from 1 to 0 across the *AI* transition, and exhibits an inflexion point at the centre of such a crossover, where $M_T = 0.5$.

Notice that the reduced modulus so defined is, in principle, affected by all dispersive effects influencing ω_s, *i.e.* not only the the negative dispersion associated with the *AI* transition, but also the competing positive sound dispersion (*PSD*) arising from the viscoelasticity of the system. The actual value of M_T should thus be explicitly corrected for the latter effects (Bencivenga *et al.*, 2006).

Figure 8-8 displays the M_T values obtained from the *IXS* spectra nitrogen (Bencivenga *et al.*, 2006), after *PSD* effect corrections and as a function of the parameter ξ.

In this modelling, the parameters defining the Rayleigh therm were kept fixed by using (macroscopic, or $Q = 0$) values tabulated in the NIST database (NIST) as well as direct measurements of $S(Q)$. In Figure 8-13 the values so obtained of M_T are reported as a function of the variable $\xi = D_T Q/c_s$. Not unexpectedly, M_T sharply

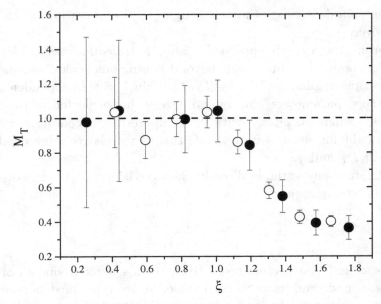

Figure 8-13:   The values of the reduced modulus M_T (see text) as derived from *IXS* measurements on nitrogen at $T = 171\,\mathrm{K}$ (open circles) and 190 K (dots) as a function of the adimensional parameter $\xi = D_T Q/c_s$ (see text). The horizontal dashed line represents the value pertinent to adiabatic regime. Data are redrawn from Ref. (Bencivenga, Cunsolo *et al.*, 2006).

decreases from its adiabatic unit value upon increasing ξ, although its transition to the isothermal regime ($M_T = 0$) is not fully captured by the experimental data in the plot. Furthermore, the apparent inflexion point of the curve seems shifted slightly upward from its expected position ($\xi = 1$).

8.9. Approximating the Rayleigh contribution to the memory function

As mentioned, most *IXS* works based on the memory function formalism used the *MH* model of the memory function defined by Eqs. (8.6–8.7).

Such a memory function contains several free model parameters to be ultimately optimized through a best-fit to the experimentally measured spectral shape. In most situations, the number of free parameters is too large to be practically manageable in the fitting procedure. Consequently, if all parameters are left free to vary without any external constraint, the χ^2 minimization routine delivers best-fit results with sizable statistical correlation. This problem becomes especially severe when modeling the lineshape of noble gases, due to the highly damped collective modes and the relatively low sound velocity of these systems. For this reason, the achievement of reliable fitting results requires further approximations or external physical constraints to the used model.

For instance, the Rayleigh contribution, *i.e.* the last exponential term in the memory function defined above, is sometimes discarded based on the assumption $\gamma(Q) \approx 1$ (Cunsolo *et al.*, 2012), which is reasonably accurate when modelling the high Q spectrum of noble gases, or Lennard–Jones systems, at sufficiently high Q's.

As an alternative option, the Rayleigh term is sometimes kept fixed to its hydrodynamic value while replacing all relevant parameters with their macroscopic counterparts (Monaco *et al.*, 1999, Bencivenga *et al.*, 2007a, Bencivenga *et al.*, 2007b), i.e. assuming:

$$\omega_0^2(Q)(\gamma(Q) - 1)\exp\left(-\gamma(Q)D_T(Q)Q^2t\right)$$
$$\equiv k_BTQ^2/[MS(Q)](\gamma - 1)\exp(-\gamma D_T Q^2 t). \qquad (8.10)$$

Handling explicitly the Q-dependence of this approximated Rayleigh term requires some knowledge of the static structure factor $S(Q)$. For instance, in their *IXS* work on water, Monaco and collaborators (Monaco *et al.*, 1999) assumed $S(Q) \approx S(0)$, based on the low Q range covered by the measurement. Conversely, Bencivenga and collaborators covered a more extended Q interval and explicitly measured $S(Q)$, using the value as an input for the best fit modeling of *IXS* lineshape.

Keeping the generalized thermal parameters fixed to their $Q = 0$ value is often a stretch which critically impacts on best-fit results. For instance, the $D_T(Q) \approx D_T$ approximation in Eq. (8.10) is a broadly used assumption which ultimately imposes a parabolic growth in the Rayleigh peak widths, thereby poorly reproducing the observed experimental trend at high Q's. In these perspectives, the following

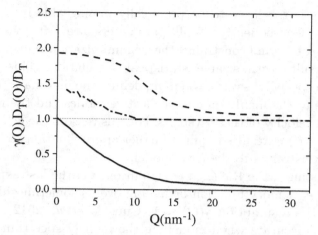

Figure 8-14:   The Q-dependence of the generalized specific heats ratio, $\gamma(Q)$ and the normalized thermal diffusivity, $D_T(Q)/D_T(0)$, are reported as a dashed and a solid line as derived for an LJ model simulating liquid Argon at the triple point (Levesque *et al.*, 1973). The plot also includes, as a dash-dotted line, the value of $\gamma(Q)$ obtained for an MD simulation on supercritical neon (Cunsolo *et al.*, 1998).

questions become of pivotal interest:

(1) To what extent is the $D_T(Q) \approx D_T$ approximation supported by literature results and inside what Q range?

(2) Given that this approximation affects the thermal contribution to the memory function, what is the estimated relative amplitude of this contribution?

   The Q-dependence of $D_T(Q)$ and $\gamma(Q)$ is reported in Figure 8-14 as derived from MD computations on LJ models simulating liquid Ar (Levesque *et al.*, 1973) and supercritical Ne (Cunsolo *et al.*, 1998).

   The plot clearly shows that the parameters defining the Rayleigh or thermal contribution to the memory function have non-negligible Q-dependence. However, the plotted curves indicate that, at sufficiently large Q, $\gamma(Q) \approx 1$, which implies that the Rayleigh vanishes at these Q values. Interestingly, this partially supports the pure viscoelastic modeling of the memory function. We recall here that the latter rests on both the single timescale and the $\gamma(Q) \approx 1$ assumptions.

A validity check for the pure viscoelastic hypothesis in a simple monatomic system is provided by an *IXS* work on liquid and supercritical Ne in which a pure viscoelastic model was used to approximate the measured lineshapes (Cunsolo *et al.*, 2001). It was observed that the strength of the thermal contribution, $\omega_0^2(Q)(\gamma(Q)-1)$, typically amounted to about 10% of the viscous strength, $[c_\infty^2(Q) - c_0^2(Q)]Q^2$.

Clearly, a more quantitative and thorough assessment of how the various approximations used for the Rayleigh contribution ultimately influence the outcome of the lineshape analysis would be highly beneficial. Aside from this much-needed check, we stress here the critical role that computer simulations could potentially play in support of experimental results when performing this critical evaluation.

Appendix 8A

We are here interested in a demonstration of the result in Eq. (8.7), which expresses the intercollision time for a system of a gas of classical hard-spheres (*HS*). To this purpose, we can start from the description of a collision event involving two *HS*, here labelled as "1" and "2". It is worth stressing that the collision does not involve a moving *HS* which hits a target *HS* at rest, but rather two *HS* approaching each other with a relative velocity:

$$v_{rel} = v_1 - v_2.$$

Correspondingly, the mean square velocity reads as:

$$\langle v_{rel}^2 \rangle = \sqrt{\langle v_1^2 \rangle - 2\langle v_1 \cdot v_2 \rangle + \langle v_2^2 \rangle} = \sqrt{2}\sqrt{\langle v^2 \rangle},$$

where in the second equality it was considered that, being the movements of the two *HS* uncorrelated, $\langle v_1 \cdot v_2 \rangle = 0$, furthermore, being the *HS* identical, $\langle v_1^2 \rangle = \langle v_2^2 \rangle = \langle v^2 \rangle$.

If σ_{HS} is the diameter of two colliding *HS*, the effective cross-section is a circular area of diameter $2\sigma_{HS}$ (see Figure 8-15) and area $\pi\sigma_{HS}^2$.

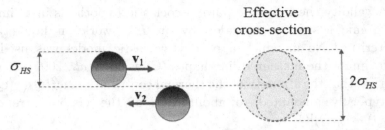

Figure 8-15:   A scheme illustrating the cross-section of two colliding hard spheres of diameter σ_{HS} (see text).

Within a time-lapse Δt, this area spans a distance

$$d(t) = \sqrt{\langle v^2 \rangle}\,\Delta t,$$

and swipes a volume:

$$V(t) = \pi \sigma_{HS}^2 \sqrt{\langle v^2 \rangle}\,\Delta t.$$

Let us now consider the number of collisions occurring in the time lapse Δt within the spanned volume $V(t)$. If the target *HS* were at rest, the number of such collisions would have been simply equal to the number of *HS* contained in the swept volume, $nV(t)$, where n is the number density of the *HS* system. However, since the target *HS* are actually moving toward the hitting ones with relative velocity $\sqrt{2}\sqrt{\langle v_2 \rangle}$, the number of collisions inside the volume $V(t)$ increases by a factor $\sqrt{2}$ respect to the fixed target scenario. We can correspondingly evaluate the mean free path as:

$$\lambda = \frac{d(t)}{\sqrt{2}nV(t)} = \frac{1}{\sqrt{2}n\pi\sigma_{HS}^2}.$$

At this stage, the mean inter-collision time τ_{coll} can be evaluated as the ratio between the mean distance traveled by the single molecule in the lapse Δt and the number of collisions occurring in the same lapse. Notice that $\langle v \rangle$ ($\neq \sqrt{\langle v^2 \rangle}$) represents the mean velocity of the single molecule, which for a gas of Boltzmann hard spheres reads as $\langle v \rangle = \sqrt{8k_B T/(\pi M)}$, as deduced by evaluating the average with a Boltzmann distribution. In summary:

$$\tau_{coll} = \frac{\lambda}{\langle v \rangle} = \frac{1}{\langle v \rangle} \frac{1}{n\sqrt{2\pi\sigma_{HS}^2}} = \frac{1}{n}\sqrt{\frac{M}{16\pi k_B T \sigma_{HS}^4}},$$

which is the same result expressed by Eq. (8.7).

References

C. A. Angell, in *Relaxations in Complex Systems*, edited by K. L. Ngai and G. B. Wright, (Naval Research Laboratory, Washington, DC, 1984). Pag. 3.

Bencivenga, F., Cunsolo, A., Krisch, M., Monaco, G., Ruocco, G. & Sette, F. 2006. *Europhys. Lett.*, **75**, 70.

Bencivenga, F., Cunsolo, A., Krisch, M., Monaco, G., Orsingher, L., Ruocco, G., Sette, F. & Vispa, A. 2007a. *Phys. Rev. Lett.*, **98**, 085501.

Bencivenga, F., Cunsolo, A., Krisch, M., Monaco, G., Ruocco, G. & Sette, F. 2007b. *Phys. Rev. E*, **75**, 051202.

Bencivenga, F., Cimatoribus, A., Gessini, A., Izzo, M. G. & Masciovecchio, C. 2009. *J. Chem. Phys.*, **131**, 144502.

Berne, B. J. & Pecora, R. 1976. *Dynamic Light Scattering*, New York, Wiley.

Bohmer, R., Hinze, G., Diezemann, G., Geil, B. & Sillescu, H. 1996. *Europhys. Lett.*, **36**, 55.

Boon, J. P. & Yip, S. 1980. *Molecular Hydrodynamics*, Dover, Courier Dover Publications.

Cohen, M. H. & Turnbull, D. 1959. *J. Chem. Phys.*, **31**, 1164.

Cummins, H., Li, G., Du, W., Pick, R. M. & Dreyfus, C. 1996. *Phys. Rev. E*, **53**, 896.

Cunsolo, A. 1999. Thesis, Universite' J. Fourier.

Cunsolo, A. 2015. *Adv. Condens. Matter Phys.*, **2015**, 137435.

Cunsolo, A., Kodituwakku, C. N., Bencivenga, F., Frontzek, M., Leu, B. M. & Said, A. H. 2012. *Phys. Rev. B*, **85**.

Cunsolo, A. & Nardone, M. 1996. *J. Chem. Phys.*, **105**, 3911.

Cunsolo, A., Pratesi, G., Ruocco, G., Sampoli, M., Sette, F., Verbeni, R., Barocchi, F., Krisch, M., Masciovecchio, C. & Nardone, M. 1998. *Phys. Rev. Lett.*, **80**, 3515.

Cunsolo, A., Pratesi, G., Verbeni, R., Colognesi, D., Masciovecchio, C., Monaco, G., Ruocco, G. & Sette, F. 2001. *J. Chem. Phys.*, **114**, 2259.

Cunsolo, A., Ruocco, G., Sette, F., Masciovecchio, C., Mermet, A., Monaco, G., Sampoli, M. & Verbeni, R. 1999. *Phys. Rev. Lett.*, **82**, 775.

Cunsolo, A., Suvorov, A. & Cai, Y. Q. 2016. *Philos. Mag.*, **96**, 732.

Davidson, D. W. & Cole, R. H. 1951. *J. Chem. Phys.*, **19**, 1484.

Davis, C. M. & Jarzynsky, J. 1972. *Water: A Comprehensive Treatise*, New York: Plenum.

Debenedetti, P. G. Metastable Liquids (Princeton University Press, Princeton, (1996)).

Dreyfus, C., Lebon, M., Cummins, H., Toulouse, J., Bonello, B. & Pick, R. 1992. *Phys. Rev. Lett.*, **69**, 3666.

Fulcher, G. S. 1925. *J. Am. Ceram. Soc.*, **8**, 339.

Giugni, A. & Cunsolo, A. 2006. *J. Phys. Condens. Matter*, **18**, 889.

Glorieux, C., Nelson, K., Hinze, G. & Fayer, M. 2002. *J. Chem. Phys.*, **116**, 3384.

Gotze, W. 1991. *In:* Hansen, J. P., Levesque, D. & Zinn-Justin, J. (eds.) *Liquids, Freezing and the Glass Transition.* Amsterdam: North-Holland.

Gotze, W. & Sjogren, L. 1992. *Rep. Prog. Phys.*, **55**, 241.

Gotze, W. & Sjogren, L. 1995. *Transp. Theory Stat. Phys.*, **24**, 801.

Herzfeld, K. F. & Litovitz, T. A. 1965. *Absorption and Dispersion of Ultrasonic Waves*, London, Academic Press.

S. Hunklinger & W. Arnold 1976. in Physical Acoustics, W.P. Mason, R.N. Thurston (Eds.), Vol. XII, Academic Press, New York, p. 155.

Jeong, Y. H. 1987. *Phys. Rev. A*, **36**, 766.

Jeong, Y. H., Nagel, S. R. & Bhattacharya, S. 1986. *Phys. Rev. A*, **34**, 602.

Kestin, J., Sengers, J. V., Kamgar-Parsi, B. & Sengers, J. M. H. L. 1984. *J. Phys. Chem. Ref. Data*, **13**, 175.

Keyes, T. & Kivelson, D. 1971. *J. Chem. Phys.*, **54**, 1786.

Laidler, K. J. 1987. *Chemical Kinetics*, Harper & Row, New York.

Levesque, D., Verlet, L. & Kurkijarvi, J. 1973. *Phys. Rev. A*, **7**, 1690.

Litovitz, T. A. 1959. *J. Acoust. Soc. Am.*, **31**, 681.

Litovitz, T. A. & McDuffie, G. E. 1963. *J. Chem. Phys.*, **39**, 729.

Macedo, P. & Litovitz, T. 1965. *J. Chem. Phys.*, **42**, 245.

Madden, P. & Impey, R. 1986. *Chem. Phys. Lett.*, **123**, 502.

Magazu, S., Maisano, G., Majolino, D., Mallamace, F., Migliardo, P., Aliotta, F. & Vasi, C. 1989. *J. Phys. Chem.*, **93**, 942.

Mazzacurati, V., Nucara, A., Ricci, M. A., Ruocco, G. & Signorelli, G. 1990. *J. Chem. Phys.*, **93**, 7767.

Monaco, A., Scopigno, T., Benassi, P., Giugni, A., Monaco, G., Nardone, M., Ruocco, G. & Sampoli, M. 2004. *J. Chem. Phys.*, **120**, 8089.

Monaco, G., Cunsolo, A., Ruocco, G. & Sette, F. 1999. *Phys. Rev. E*, **60**, 5505.

Monaco, G., Masciovecchio, C., Ruocco, G. & Sette, F. 1998. *Phys. Rev. Lett.*, **80**, 2161.

Montrose, C. J. & Litovitz, T. A. 1970. *J. Acoust. Soc. Am.*, **47**, 1250.

Mountain, R. D. 1966. *Rev. Mod. Phys.*, **38**, 205.

NIST. *Thermodynamic data are from the database*: http://webbook.nist. gov/chemistry/form-ser.html.

Phillips, J. C. 1996. *Rep. Prog. Phys.*, **59**, 1133.

Piccirelli, R. & Litovitz, T. 1957. *J. Acoust. Soc. Am.*, **29**, 1009.

Pinnow, D. A., Candau, S. J., LaMacchia, J. T. & Litovitz, T. A. 1968. *J. Acoust. Soc. Am.*, **43**, 131.

Pontecorvo, E., Krisch, M., Cunsolo, A., Monaco, G., Mermet, A., Verbeni, R., Sette, F. & Ruocco, G. 2005. *Phys. Rev. E*, **71**, 011501.

Pratesi, G., Bellosi, A. & Barocchi, F. 2000. *Eur. Phys. J. B*, **18**, 283.

Rouch, J. 1976. *J. Chem. Phys.*, **65**, 4016.

Rouch, J., Lai, C. C. & Chen, S.-H. 1977. *J. Chem. Phys.*, **66**, 5031.

Ruocco, G., Sette, F., Di Leonardo, R., Fioretto, D., Krisch, M., Lorenzen, M., Masciovecchio, C., Monaco, G., Pignon, F. & Scopigno, T. 1999. *Phys. Rev. Lett.*, **83**, 5583.

Ruocco, G., Sette, F., Di Leonardo, R., Monaco, G., Sampoli, M., Scopigno, T. & Viliani, G. 2000. *Phys. Rev. Lett.*, **84**, 5788.

Scopigno, T., Balucani, U., Ruocco, G. & Sette, F. 2000. *Phys. Rev. Lett.*, **85**, 4076.

Scopigno, T., Ruocco, G. & Sette, F. 2005. *Rev. Mod. Phys.*, **77**, 881.

Sette, F. Krisch M. H., Masciovecchio, C., Ruocco, G., Monaco, G. & Sette, F. 1998. *Science*, **280**, 1550.

Sette, F., Ruocco, G., Krisch, M., Masciovecchio, C., Verbeni, R. & Bergmann, U. 1996. *Phys. Rev. Lett.*, **77**, 83.

Slie, W. M., Donfor, A. R. & Litovitz, T. A. 1966. *J. Chem. Phys.*, **44**, 3712.

Teixeira, J., Bellissent-Funel, M. C., Chen, S. H. & Dorner, B. 1985. *Phys. Rev. Lett.*, **54**, 2681.

Torre, R. 2007. *Time-Resolved Spectroscopy in Complex Liquids*, Springer, New York.

Torre, R., Bartolini, P. & Righini, R. 2004. *Nature*, **428**, 296.

Trinh, E. & Apfel, R. E. 1980. *J. Chem. Phys.*, **72**, 6731.

Walrafen, G. E. 1972. in *Water: A Comprehensive Treatise*. Editor: F. Franks, Plenum Press, New York, Vol. 1, pag. 151.

Williams, G. & Watts, D. C. 1970. *Trans. Faraday Soc.*, **66**, 80.

Zwanzig, R. 1965. *J. Chem. Phys.*, **43**, 714.

Chapter 9

A few emerging, controversial or unsolved topics in *IXS* investigations of simple fluids

9.1. How do relaxation processes depend on thermodynamic conditions?

An increasing number of *IXS* studies of high-frequency relaxation phenomena in amorphous materials focus on the search for universal, rather than distinctive, phenomenological aspects. The general aim of these studies is to identify some trends shared by diverse systems which unveil fundamental dynamic properties of all fluids, or, generally, disordered systems.

Along these lines, it is worth mentioning a comprehensive *IXS* investigation (Bencivenga *et al.*, 2007) on relaxation phenomena in a few prototypical samples having an increasing microscopic complexity: a monatomic (neon), a diatomic (nitrogen) and an associated fluid (ammonia), as well as an associated fluid having unexplained thermodynamic anomalies, such as liquid water. In all cases, the line-shape model was derived within the Mori–Zwanzig formalism based on a second-order *MH* memory function model, as the one defined in Eq. (8.8); in this model, all parameters relevant to the thermal contribution were kept fixed (see Eq. (8.9)) to the value derived from the literature. In order to estimate the macroscopic $(Q = 0)$ values of relaxation parameters derived from the best-fit of measured lineshape, their Q-dependence was modelled with an empirical exponential law.

As an example, the $Q = 0$ extrapolated values of the compliance relaxation time reported by Bencivenga and collaborators

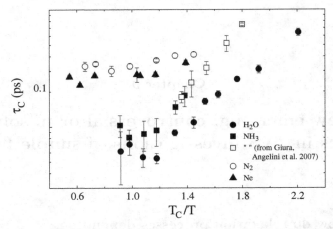

Figure 9-1:   The $Q = 0$ extrapolated relaxation timescale measured by *IXS* in Ref. (Bencivenga, Cunsolo *et al.* 2009) for several prototypical fluids, as shown in the legend. The relaxation times of ammonia extracted from the *IXS* work in Ref. (Giura, Angelini *et al.*, 2007) are also reported for comparison as open squares.

are displayed in Figure 9-1 as a function of the dimensionless variable T_c/T, where the suffix "c" labels the value at the critical point. It appears that the relaxation timescale of non-associated systems (neon and nitrogen) has a mild dependence, if any, on temperature, while for associated ones (water and ammonia) such dependence is more pronounced, at least at subcritical temperatures. This trend suggests that in the liquid phase, the dominant relaxation processes in associated liquids acquire a predominantly structural character.

However, a fair comparison between the dominant timescale of different systems requires normalization for a suitable reference variable, which in the *IXS* work of Ref. (Bencivenga *et al.*, 2007) was identified with the intercollision time (see Eq. (8.7)).

To get an idea on the possible connection between relaxation and intercollision timescales, it is useful to refer to Figure 9-2, which displays the ratio $\psi_C = \tau_C/\tau_{\text{coll}}$ as a function of the reduced variable T_c/T, where the shorthand notation $\tau_C = \tau_C(Q = 0)$ indicates the macroscopic extrapolation of the compliance relaxation time.

The comparison with Figure 9-1 shows that this rescaling forces data to merge into some master curve, which, in the temperature range covered by data, can be reasonably well approximated by two distinct linear segments. This behaviour prompts two main conclusions:

(1) The flat T-dependence at nearly critical and supercritical conditions mirrors the predominance of collisional relaxation phenomena, which, for hydrogen-bonded materials (water and ammonia), likely stems from the weakening of the hydrogen bond (HB) network at supercritical temperatures. Indeed, at these temperatures, the dominant relaxation time seemingly parallels the intercollision time of a hard sphere gas.

(2) Conversely, the steeper T-dependence at subcritical conditions unveils the emergence of structural relaxation processes in these associated systems.

In summary, data in Figure 9-2 suggest that structural and collision relaxations are two prototypical relaxation phenomena dominating either in interconnected systems or in gas-like ones, respectively.

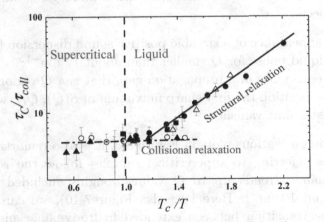

Figure 9-2: The data in Figure 9-1 are here rescaled for the corresponding collisional relaxation time (see text). The straight lines through the data serve as guide-to-eye to emphasize the two distinct temperature behaviours. The vertical dashed line locates the critical temperature.

9.2. Liquid-like and compressed gas behaviour

Figure 9-2 emphasizes distinctive aspects of the relaxational behaviours of liquid and supercritical systems. The outlined trend raises natural questions on whether, and to what extent, these differences are reflected by the Q-dependence of transport parameters relevant for the spectral shape. To partially address this point, the upper panels of Figure 9-3 compare the dispersion curve of:

(1) **Supercritical Ne at (T = 294 K, P = 3 Kbar) and He at (T = 25.5 K, P = 0.8 Kbar)** as derived from an MD simulation (Cunsolo *et al.*, 1998) and an *IXS* measurement (Verbeni *et al.*, 2001, Cunsolo, 1999), respectively. To give a reference, the respective critical points of the two systems are (T_c = 5.2 K, P_c = 2.26 bar) and (T_c = 44.5 K, P_c = 27.7 bar).

(2) **Liquid Ne at (T = 35 K, P = 80 bar, n = 33.4 atoms/nm^3) and He at (T = 4 K, P = 1 bar, n = 19.5 atoms/nm^3)** as derived in Refs. (van Well and de Graaf, 1985) and (Crevecoeur *et al.*, 1995), respectively.

Overall, the sound dispersion curves (two top panels of Figure 9-3) in these liquid and supercritical noble gases differ in two main aspects:

(1) The appearance of a sizeable positive sound dispersion (PSD) in the liquid phase for Q smaller than about 12 nm^{-1};

(2) The emergence of a propagation gap, that is a Q region centred at the position the first sharp maximum of $S(Q)$, Q_m, where the inelastic shift vanishes.

Looking at various plots in Figure 8-3, it is also remarkable that all curves referring to supercritical samples follow the respective hydrodynamic trends up to Q values roughly included between 5 nm^{-1} and 10 nm^{-1}. Here (see also Figure7-10), we can roughly locate the transition between extended hydrodynamic and kinetic regime in these supercritical systems.

Both *PSD* and propagation gap appear even more pronounced in the dispersion curves of Figure 9-4 which are extracted from

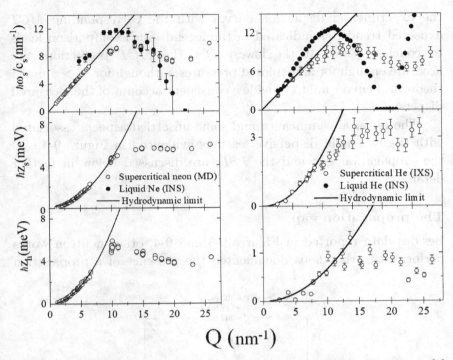

Figure 9-3:   Best-fit shape parameters of the Generalized Hydrodynamic model in Eq. (6.6) (open circles) in supercritical neon (left column) and helium (right column). Data refer to the results of *MD* simulations and *IXS* measurements in Ref. (Cunsolo, Pratesi *et al.* 1998) and (Cunsolo 1999), respectively. Dispersion curves measured by INS in liquid neon (van Well and de Graaf 1985) and helium (Crevecoeur, Verberg *et al.* 1995) are also reported for comparison (dots) in the left and right plots on the top, respectively. In each plot, the solid lines represent the hydrodynamic prediction derived by inserting in Eqs. 6.2 transport parameters reported in the NIST database (NIST) and bulk viscosity data from Ref. (Castillo and Castaneda 1988).

an *INS* work on liquid Ar at different pressures by de Schepper and collaborators (de Schepper *et al.*, 1984b). In this work, the authors attempted to describe the *PSD* using the mode-coupling theory (*MCT*) prediction (Gotze, 2008), according to which such dispersive behaviour can be expressed by a truncated Q-expansion with fractional exponents. The accuracy of this prediction can be judged from Figure 9-4 by comparing the low Q portion of the experimental dispersion curves with the corresponding *MCT* expected trends, as indicated in the legend. Such comparison leads to conclude that, at the lowest *Q*'s, the *MCT* predictions are accurate enough for all explored pressures, although for $Q > 4 \, \text{nm}^{-1}$ theoretical curves fail to provide a consistent account of the measured dispersion.

The physical significance and some unsettled aspects associated with the two main dispersive features evidenced in Figure 9-4, *i.e.* the propagation gap and the *PSD*, are discussed below in further detail.

The propagation gap

Besides data reported in Figures 9-3 and 9-4, other neutron works performed in the 1980s documented the presence of a propagation

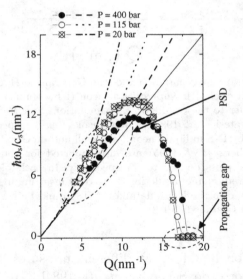

Figure 9-4:   The sound dispersion curves measured in liquid Ar at increasing pressure (de Schepper *et al.*, 1984b) and normalized to the respective sound velocities, as derived from thermodynamic data (NIST). The arrow indicates the presence of a propagation gap, i.e. a *Q* region where the sound frequency vanishes. The solid line represents the hydrodynamic limiting dispersion expectedly joined by all curves in the $Q = 0$ limit. The dashed curves are the low *Q* prediction of the Mode Coupling Theory as derived for each datum according to the legend.

gap in liquid argon (de Schepper *et al.*, 1984c, Verkerk *et al.*, 1987) and in a molten salt (McGreevy and Mitchell, 1985); more recent examples of the same effect can be found, for instance, in a computer simulation of the molecular centers of mass dynamics of methane (Sampoli *et al.*, 2008).

The kinetic theory (KT) approach (see Section 7.4 and Appendix 7A) does predict the occurrence of a propagation gap (Cohen *et al.*, 1984). Cohen and collaborators investigated this effect in the framework of the generalized Enskog theory and interpreted it as the result of a balance between competing dynamic effects. For sufficiently small Q's, elastic forces prevail, thus enabling the propagation of sound waves. At larger Q's, dissipative forces become dominating to the point of preventing the sound mode from propagating in a limited Q-window around the position first $S(Q)$ maximum, Q_m. At even higher Q values lying beyond the position of the first diffraction peak, the free streaming term of the Enskog operator (see Appendix 7A) has increasing relevance, eventually overwhelming dissipative forces and restoring a density wave propagation.

However, the actual physical significance of the gap originated some debate in the past, since it occurs around Q_m, where, owing to the overall slowdown of collective dynamics and the de Gennes narrowing, the spectrum reduces to a narrow featureless peak. For this reason, the use of the triple-peaked models yielding the propagation gap (as the GH model in Eq. (6.6)) in this Q-range was questioned (Lovesey, 1984). However, this choice was justified by recognizing that a similar lineshape profile is also consistent with a kinetic theory (KT) description of the spectrum as a sum of generalized Lorentzian terms (see Appendix 7A); indeed, it was experimentally demonstrated that the three leading terms of such an expansion yield a shape isomorph to the GH one, which is adequate to consistently describe the INS spectrum (de Schepper *et al.*, 1984a).

The positive sound dispersion

The second remarkable feature in the dispersion curve of many liquids is a sizable PSD, as often discussed throughout this book, and interpreted as the fingerprint of a viscoelastic behavior. In spite of the frequent observation of this effect, it is still unclear why it is more pronounced in some systems and how it depends on thermodynamic conditions. As mentioned, a conspicuous *PSD* has been sometimes associated with a large Einstein frequency (see Eq. (2.21)) and it was proposed (see Balucani and Zoppi, 1994) that its amplitude is estimated by the parameter $\Gamma_{PD} = M\Omega_0^2 d_0^2/(k_B T)$ where d_o is the distance at which the interaction potential reaches a minimum. This prediction applies to systems of classical particles interacting via a potential of the form $U(r) = (\varepsilon/a - b)[b(d_o/r)^a - a(d_o/r)^b]$ (whose minimum is $U(d_o) = -\varepsilon$).

In particular, based upon computer simulation of Levesque and Verlet (Levesque and Verlet 1970) on Lennard–Jones systems — for which $a = 12, b = 6$ — it can be deduced that an approximated form of the Einstein frequency is given by $T^{1/2}n^2ab$ (Lewis and Lovesey 1977). Unfortunately, using this formula to derive the expected PSD amplitude Γ_{PD} fails to provide an accurate prediction of the dispersive trends shown in Figure 9-3. Overall, a quantitative understanding of *PSD* effects still represents a challenge.

Further evidence of the viscoelastic behaviour of transport parameters

Figure 9-5 compares two spectral parameters defining the extended acoustic mode of Ne, i.e. the inelastic shift and damping, as derived for neon from *IXS* measurements and *MD* simulations in the supercritical phase (Cunsolo et al., 1998, Cunsolo, 1999) and *INS* measurements (van Well and de Graaf, 1985) on a liquid sample. Overall, data reported in the figure indicate that, while in the liquid phase the low Q values of sound frequency and damping are respectively higher and lower than heir hydrodynamic counterparts, no substantial discrepancy is instead observed in supercritical conditions.

As discussed, the dispersive effects observed in the liquid sample can be ascribed to the onset of a relaxation process, as the

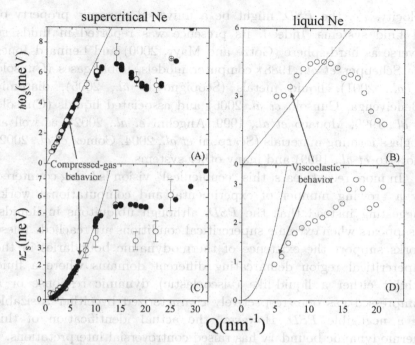

Figure 9-5:   Left column: shift (panel a) and half-width (panel c) of the inelastic peak in the spectrum of supercritical Ne at $T = 294$ K and $n = 29$ atoms/nm^3 as derived from a best-fit line-shape analysis performed with the model in Eq. (6.6). Open circles and dots are *IXS* and *MD* data from Refs. (Cunsolo, 1999) and (Cunsolo *et al.*, 1998), respectively. The solid lines represent the corresponding simple hydrodynamic prediction. Right column: the corresponding quantities are reported for liquid neon at $T = 35$ K and $n = 36.65$ atoms/nm^3. These data are adapted from Ref. (van Well and de Graaf, 1985).

viscous-to-elastic transition is revealed by an increase of sound velocity and a decrease of damping (see Section 6.2.4).

We will hereafter quote as "liquid-like" a fluid exhibiting a sizable viscoelastic behaviour and "compressed gas-like" a fluid whose dynamic response bears no evidence of it.

9.3. Evidence of (thermo)dynamic boundaries

A large body of experimental and computational results have supported the notion that the viscoelastic behaviour of the sound velocity, *i.e.* the *PSD*, might be a universal dynamic property of all fluid systems. Indeed, its presence was reported in fluids as diverse as hard-sphere (Gotze and Mayr, 2000) and Lennard-Jones (de Schepper *et al.*, 1988) computer models, noble gases (Cunsolo *et al.*, 2001), liquid metals (Scopigno *et al.*, 2005), diatomic (Bencivenga, Cunsolo *et al.* 2006) and associated liquids (Cunsolo *et al.*, 1999, Monaco *et al.*, 1999, Angelini *et al.*, 2002), as well as in glass-forming materials (Scarponi *et al.*, 2004, Comez *et al.*, 2002, Fioretto *et al.*, 1999) and many other systems.

In more recent years, this "ecumenical" vision is being countered by a growing number of experimental and computational works suggesting instead that the *PSD*, although ubiquitous in liquids, disappears when extreme supercritical conditions are reached. These works support the existence of thermodynamic boundaries in the supercritical region demarcating different domains where a fluid exhibit either a liquid-like (viscoelastic) dynamic response or a compressed gas one, respectively characterized by either a sizable or a negligible *PSD*. However, the actual identification of this thermodynamic boundary has raised controversial interpretations.

Experimental insight into this topic requires thorough investigations of the terahertz spectrum of supercritical fluids at extreme P and T values. For many decades, this has been an almost prohibitive task, owing to significant technical difficulties in reaching extreme thermodynamic conditions, due to the large samples typically required for terahertz inelastic measurements.

Advances in the X-Ray focusing and the consequently smaller focal spots of X-Ray beams made possible *IXS* experiments on smaller samples, while the improved performance of undulators and crystal optics have dramatically enhanced the typical *IXS* count rates. These technical improvements paved the way toward a whole new class of *IXS* investigations on samples at extreme pressures, based on the use of diamond anvil cells (*DACs*, see, *e.g.* Jayaraman, 1986).

Along this line, an *IXS DAC* measurement was performed on deeply supercritical oxygen at temperatures nearly doubling T_c and pressures exceeding P_c by more than two orders of magnitude

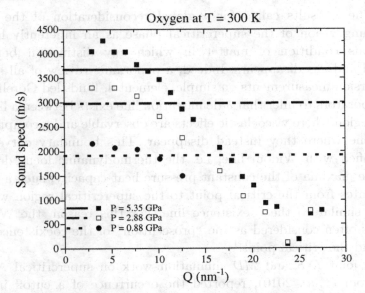

Figure 9-6: The Q-generalized sound speed derived from the IXS spectra of supercritical Oxygen at the extremely high pressures indicated in the legend. The presence of a sizable positive sound dispersion can be readily inferred from the evidence of a low Q ($\lesssim 10\,\text{nm}^{-1}$) sound speeds clearly exceeding the respective hydrodynamic predictions (horizontal lines).

(Gorelli *et al.*, 2006). Figure 9-6 displays the sound velocity extracted from these *IXS* measurements and compares it with the adiabatic sound velocity, expectedly joined in the hydrodynamic limit. The comparison with the hydrodynamic limit emphasizes the presence of a *PSD* up to about 20% in amplitude, which seems to increase upon increasing the pressure.

The evidence that the *PSD* survives in such a remote region of the supercritical domain was at odds with what previously observed in deeply supercritical neon (Cunsolo et al., 2001), *i.e.* with the anticipated disappearance of the *PSD* in supercritical fluids, as inferred from Figure 9-3. This circumstance suggested that the dynamic response in the supercritical phase is not homogeneous, yet partitioned in isolated domains in which either viscoelastic or gas-like behaviors become observable.

These results called for a global reconsideration of the long-standing vision of the supercritical phase as an inherently homogeneous condition of matter in which any distinction between liquid and gas disappears. Indeed, a panoramic review of all sound dispersion measurements on simple, elemental, fluids led Gorelli and collaborators to infer that a demarcation line exists between a liquid-like region where viscoelastic effects are observable and a compressed gas one where they instead disappear. This boundary curve was identified with Widom line, i.e. the thermodynamic locus defined by the maxima of the constant pressure heat capacity; such a locus emanates from the critical point to the supercritical region with a slope similar to the coexistence line. For this reason, the Widom line is often considered as the "prosecution" of the coexistence line beyond the critical point.[1]

A joint *IXS* and *MD* simulation work on supercritical Argon (Simeoni *et al.*, 2010), reported the occurrence of a cutoff in the *PSD* amplitude for pressures lower than the one corresponding to the Widom line (see Figure 9-7). This finding further supported the interpretation of the Widom line as a boundary between a 'liquid-like' and a 'gas-like' region, also providing a rationale for previous diffraction studies, leading to infer that the crossing of the Widom line transforms structural properties as well (Santoro and Gorelli 2008).

It is essential to recognize that any thermodynamic line demarcating a crossover between a liquid-like and gas-like dynamical response of supercritical fluids must emanate from the critical point or possibly merge into the coexistence. Otherwise, its possible penetration into the regular liquid or vapour phase would bring about the evident paradox of a partition between liquid-like and gas-like domains inside a well-defined (liquid or vapor) aggregation phase. Being the Widom line a prosecution of the coexistence line beyond

---

[1]Sometimes such thermodynamic locus is identified instead with the locus of compressibility maxima, which also emanates from the critical point (where the compressibility diverges) and is somehow close to the locus of specific heat maxima.

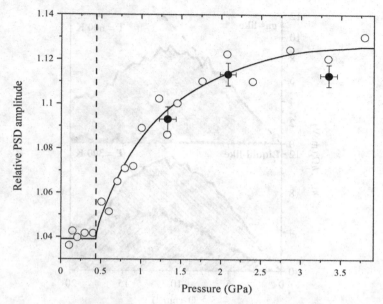

Figure 9-7: The pressure dependence of the relative positive sound dispersion amplitude in Argon at 573 K (Simeoni *et al.*, 2010), evaluated as the maximum of $c_s(Q)/c_s$. Open circles and dots represent MD simulation and IXS results, respectively. The vertical dotted line demarcates the Widom line boundary, while the two solid lines are guides to the eye separately best-fitting data before and after the Widom line. The figure is redrawn from (Simeoni *et al.*, 2010).

the critical point, it in principle represented a suitable candidate as a demarcation line.

A joint *IXS* and *MD* simulation work on deeply supercritical argon (Bolmatov, Zhernenkov *et al.* 2015) suggested the existence of a close link between a shear mode propagation and the presence of a sizeable *PSD*. In this study, *IXS* spectra were measured along an isobaric path (with $P = 1$ GPa) with T spanning the 300 K-436 K temperature range, which further extended to 800 K in the MD simulation. The analysis of the spectral shape suggested that the disappearance of the *PSD* in deeply supercritical conditions is accompanied by a parallel shear mode overdamping, as indicated by the vanishing inelastic shift of the transverse mode at high temperatures (see Figure 9-8).

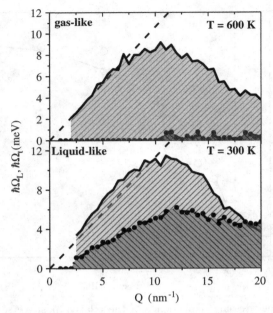

Figure 9-8:   The dispersion curves of deeply supercritical Ar at $P = 1$ GPa and the T values indicated in the plots, are reported as evaluated from the maxima positions of MD-simulated longitudinal (solid lines) and transverse (dots) current spectra. The corresponding hydrodynamic dispersions are also reported as dashed lines for comparison. Data in the figure are adapted from Ref. (Bolmatov *et al.*, 2015a).

This trend confirms the existence of distinct supercritical subdomains characterized by a different dynamic response: a gas-like region and a liquid-like one. In the former regime, viscoelastic effects are absent, and the fluid cannot support transverse wave propagation.

9.4. The Frenkel line

Some literature works challenged the interpretation of the Widom line as a dynamic crossover by showing that the extrema of different thermodynamic functions rapidly smear out above the critical point, which would make the Widom line ill-defined (Brazhkin *et al.*, 2012).

According to this alternative interpretation scheme, the thermodynamic line separating the liquid-like and the compressed gas thermodynamic regions is to be identified instead with the Frenkel line (Bolmatov *et al.*, 2013, Brazhkin *et al.*, 2012).

This thermodynamic curve was named after Y. Frenkel, who first conjectured its existence (Frenkel, 1947), and is defined as the locus of states in which the relaxation time equals the minimum period of the transverse acoustic mode. More intuitively, a "single-particle" definition of the Frenkel line makes use of a new timescale τ_F, *i.e.* the time required for a particle to jump over a distance comparable to its own size a. The other relevant timescale of the Frenkel model is the relaxation time τ, which gives a measure of the time it takes for the atoms forming the cage around the tagged particle to rearrange their positions entirely.

If τ is much longer than τ_F, the behavior of the fluid reminds the one of a solid, consisting of confined oscillations of single atoms around their equilibrium lattice positions.

Conversely, in the $\tau_F \approx \tau$ regime, rearrangements of the first neighbouring atomic cage can accommodate out-of-cage jumps of the single atom, and the system exhibits a viscoelastic behavior.

If such a (cage) relaxation is, instead, extremely rapid ($\tau \gg \tau_F$), the system dynamics reminds the one of a gas.

This approach lends itself to the introduction of the "Frenkel frequency" $\omega_F = 2\pi/\tau_F$ to be considered as an indicator of the fluid rigidity, which demarcates a low-frequency cutoff for transverse mode propagation to occur.

It was initially conjectured that this line does not define a crossover in the liquid structure, only influencing viscoelastic transport properties, as manifested through the onset of a shear wave propagation. The identification of the Frenkel line with a crossover of the first neighbour atomic arrangement was, instead, supported by joint X-Ray diffraction and MD results which demonstrated visible changes in the pair collection function upon crossing the Frenkel line (Bolmatov *et al.*, 2015b).

Despite an undoubted success in accounting for some experimental evidence, the "Frenkel line" explanation has some inherent weaknesses, as evidenced by some *IXS* results explicitly questioning this interpretetive scheme (see, e.g., Bryk *et al.*, 2017). Weak aspects can be summarized as follows:

(1) First, this model does not provide specific information on the nature of microscopic interactions and how these affect shear wave propagation.

(2) Also, the Frenkel line does not emanate from the critical point, yet it crosses the liquid region of the phase diagram. Obviously, its interpretation as a partition line between liquid-like and gas-like dynamic domains cannot be extended to this subcritical region.

(3) Finally, the Frenkel line theory predicts the existence of a cutoff frequency for shear propagation, while theories and experiments in the literature seem to support, instead, the existence of a wavelength (or Q) threshold.

9.5. To what extent does the dynamics of a disordered system resemble the one of a solid?

As opposite to viscoelastic approaches, emphasizing the resemblance between liquid and solid systems, kinetic theory approaches discussed in Chapter 7 stress the analogy between liquid and gaseous ones. However, this analogy is overestimated in many aspects, as some facts unambiguously suggest similarities between the liquid and the solid aggregation phases and, conversely, visible discontinuities between liquid and gaseous ones (Frenkel, 1947).

A striking example is provided by the density evolution across the phase transition: although a significant jump in density occurs while crossing the liquid-vapour coexistence line, a relatively small variation is usually associated with the melting. Also, the latent heat of fusion is typically several orders of magnitude smaller than the latent heat of vaporization, thus indicating that cohesive forces change upon melting much less than they do during evaporation, which is also reflected in the correspondingly smaller variation of interatomic distances (typically a few per cents). Finally, the specific heat of materials is only slightly affected across the melting.

Other aspects of the similarity between liquids and solids relate to the existence of a local order, which, in a liquid, gives rise to diffraction pattern featured by a sharp main peak; in the crystal the order becomes global and the diffraction pattern is correspondingly featured by Bragg spots. Conversely, no relic of first neighbours ordering can be found in the intensity diffracted from a dilute gas.

Further below, we aim at addressing an important question: how does the terahertz dynamic response reflect the differences in the structure of a crystalline and an amorphous material?

9.5.1. *Sound damping, structural disorder and elastic anisotropy*

Atomic movements in liquids are not bound to "lattice site" oscillations, as they are in crystals; indeed, in these systems rearrangements continuously occur, taking the form of either diffusive motions or relaxations. Although these dynamical processes provide a sizable contribution to the damping of acoustic excitations in the liquid, it is reasonable to assume that they are relevant for frequencies lower than those typical of the elastic regime. Indeed, at these high frequencies diffusions and relaxations are mostly frozen and the dominant microscopic mechanisms significantly impacting on acoustic propagation primarily arise from static properties of a liquid, i.e. from its structural arrangement. On a general ground, the structure of a liquid lends itself to two complementary representations. It can be alternatively depicted as a continuous disordered network, or , instead, as an ensemble of micro-crystallites, small enough to elude detection by ordinary diffraction methods. These microdomains are randomly oriented, but preserve an internal (local) ordering. In a scattering measurement, the momentum transfer has finite projections along the multiple crystallographic directions pertaining to these microdomains and, due to the dynamic anisotropy, density fluctuations propagate with different sound velocities along these directions. Therefore, sound waves with slightly different propagation speeds participate in the dominant acoustic mode of a liquid, and their superposition ultimately contributes to the acoustic damping.

Consistently, a portion of the line broadening of an acoustic mode measured in a liquid can be assigned to the fluid's elastic anisotropy. Expectedly, this broadening is similar to the one measurable in an "orientationally averaged" crystal, that is a system, as a polycrystal, where the crystalline domains are randomly oriented.

On the other hand, the structure of a liquid is also characterized by an inherent structural disorder, *i.e.* by the existence of a distribution of nearest neighbor distances in the spatial arrangement of the molecules. This distribution of inherent structures is mirrored by the finite width Δ of the first sharp peak of $S(Q)$. Consequently, the dominant acoustic mode measured in a liquid contains a multiplicity of sound modes pertaining to local structures having different first neighbor distances. This modulation of the molecular packing (local density) is reflected by a corresponding spread of the sound velocities ultimately contributing to the dominant acoustic excitation. It can thus be expected that the more disordered is the system, the wider the band of acoustic frequencies underlying a given excitation. In summary, the existence of a Q-spread in the pseudo-Brillouin zones of a liquid ultimately gives rise to a band of allowed dispersion curves. Again, this inherent disorder ultimately results in an additional spectral broadening of the dominating acoustic mode.

To frame the problem in a more quantitative ground, we might refer to the simple model illustrated in two IXS works on liquid metals (Giordano and Monaco, 2010, Giordano and Monaco, 2011). In these works, Giordano and Monaco have proposed a simple, first approximation, estimate of the two extremes of the dispersion band arising from the spread of pseudo-Brilluoin of a liquid. Within their model, the dispersion curve is represented as a function of the reduced wavevector Q/Q_m where Q_m is the position of the first diffraction maximum. Highest and lowest values of the dispersion band allowed for a liquid can be respectively estimated as

$$\omega_s^H \left(Q/(Q_m - \Delta/2) \right)$$

and

$$\omega_s^L \left(Q/(Q_m + \Delta/2) \right),$$

where $\omega_s^H(Q)$ and $\omega_s^L(Q)$, are respectively, the highest and lowest frequency phonon dispersions measurable in the single crystal along the various crystallographic directions. Compared to the "standard" representation in terms of the reduced vector (Q/Q_m), the above rescaling of $\omega_s^H(Q)$ and $\omega_s^L(Q)$ exacerbates their difference. In a first approximation, all dispersion curves coexisting in the liquid are expectedly included within the band $\omega_s^H(Q/(Q_m - \Delta/2))$-$\omega_s^L(Q/(Q_m + \Delta/2))$. As suggested by Giordano and Monaco, this justifies the following rough estimate of the acoustic mode of a liquid:

$$\Delta\omega_s\left(\frac{Q}{Q_m}\right) = \omega_s^H\left(\frac{Q}{Q_m - \Delta/2}\right) - \omega_s^L\left(\frac{Q}{Q_m + \Delta/2}\right).$$

In the limit of low Δ/Q_m values, this difference can be approximated by the following linear expansion:

$$\Delta\omega_s\left(\frac{Q}{Q_m}\right) \approx \omega_s^H\left(\frac{Q}{Q_m}\right) - \omega_s^L\left(\frac{Q}{Q_m}\right)$$
$$+ \frac{\Delta}{2Q_m}\left[\omega_s^H\left(\frac{Q}{Q_m}\right) + \omega_s^L\left(\frac{Q}{Q_m}\right)\right].$$

The right-hand side of the above equation is the sum of two distinct terms:

(1) A term proportional to Δ, which directly relates to the disorder-induced spread of the pseudo-Brillouin zones of the liquid.
(2) The remaining contribution, *i.e.* the difference between highest and lowest frequency single crystal's dispersions in the reduced vector representation. As discussed, this "structural disorder"-free term relates to the local elastic anisotropy, and stems from the superposition of the multiple orientations of the local "microcrystallites" composing the liquid structure.

In the IXS work of Ref. (Giordano and Monaco, 2011) it was demonstrated that the above formula, despite its inherent simplicity, provides a consistent account of the acoustic damping in of liquid Ga.

Also, in an IXS work on Na (Giordano and Monaco, 2010) the disorder-induced contribution to the damping of collective modes was singled out by comparing the *IXS* spectra from the sample in the liquid and in the polycrystalline phases. The spectral shape was described using the following model profile:

$$S\left(Q,\omega\right) = \frac{\Gamma_0}{\omega^2 + \Gamma_0^2} + S_{\text{in}}\left(Q,\omega\right) \otimes G\left(\omega\right). \tag{9.1}$$

In this model, the effect of local structures distribution is accounted for by assuming a Lorentzian broadening of the central peak and a "blurring" of the side peaks in the form of a convolution (the "$\otimes$" symbol) between an inherent crystal-like spectrum, $S_{in}(Q,\omega)$, and a Gaussian broadening, $G(\omega)$.

A similar model was used to describe the spectrum of the liquid sample only, since, for the polycrystalline one, the lack of disorder-induced broadening advised the use of a $\delta(\omega)$ function approximation, for both $G(\omega)$ and the Lorentzian profile in Eq. (9.1). The structural disorder-free, or inherent, spectrum $S_{\text{in}}(Q,\omega)$, was described by a model including two *DHO* profiles (see Appendix 6A), accounting for the coexistence of a longitudinal and a transverse mode.

Interestingly, the outcome of this modeling indicates that $S_{\text{in}}(Q,\omega)$ is essentially the same for crystal and liquid samples, which supports the interpretation of this profile as a "disorder free" spectral density.

The main topics discussed above are summarized by the results of Giordano and Monaco (Giordano and Monaco, 2010) displayed in Figure 9-9.

There, the dispersion curves of Na obtained from best fitting the model in Eq. (9.1) to the *IXS* spectra of liquid Na are reported as a function of the reduced momentum Q/Q_m. Error bars $\pm z_s(Q)$ are assigned to each point, with $z_s(Q)$ being the damping of the *DHO* component of the spectrum, $S_{in}(Q,\omega)$; here, $z_s(Q)$ represents the component of the damping independent on the structural disorder, which turned out to be similar for the liquid and the polycrystalline samples. The plot includes the error band defining the overall broadening of the acoustic excitation in the liquid, defined as $\omega_s(Q) \pm \Gamma_s(Q)$, with $\Gamma_s(Q) = z_s(Q) + W_G(Q)$ being the total spectral damping and $W_G(Q)$ the halfwidth of the Gaussian broadening function $G(\omega)$. Figure 9-9 includes the highest frequency ([1 1 0])

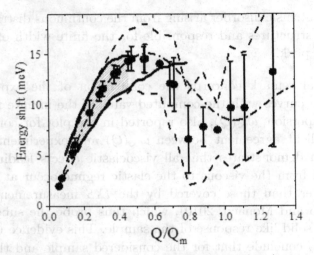

Figure 9-9: The sound dispersion curve of liquid Na (dots) is reported along with the highest frequency and lowest frequency acoustic branches measured in the single crystal plotted as a function of $Q/Q_m - \Delta/2$ and $Q/Q_m - \Delta/2$, respectively (thick solid lines). The error bars associated with each point are $\pm z_s(Q)$ where $z_s(Q)$ is the width of the inelastic mode of the $S_{in}(Q,\omega)$ profile in Eq. (9.1). The two dashed lines delimiting the $\omega_s(Q) \pm \Gamma_s(Q)$ are also reported for reference (see text). Finally, the value of $\omega_\infty(Q)$ computed for the liquid sample is reported for comparison as a thick dash-dotted line. All curves are redrawn from Ref. (Giordano and Monaco, 2010).

and lowest frequency ([1 0 0]) acoustic branches measured in the single crystal $\omega_s^H Q/Q_m - \Delta/2$ and $\omega_s^L Q/Q_m + \Delta/2$, respectively. As discussed, these two phonon branches demarcate lower and upper limits of the dispersion measurable in the system, and their difference can be taken as a measure of the sound dispersion spread (damping) arising from the continuous distribution of structures of the liquid. Notably, such a spread is consistent with the overall acoustic broadening of the liquid within the overlapping range.

Overall, data in Figure 9-9 confirm the conclusion that two main factors determine the linewidth of a propagating collective mode in a liquid. Namely:

(1) The random orientation of the locally ordered microdomains, which yields a similar contribution in the liquid and polycrystalline samples.

(2) The intrinsic disorder arising from the continuous distribution of local structures and responsible for the finite width of the first $S(Q)$ peak.

A final remark concerns the comparison of the experimental dispersion curves with the computed value of the infinite frequency sound dispersion, $\omega_\infty(Q)$, also reported in the plot for comparison. The excellent agreement between $\omega_\infty(Q)$ and experimental sound dispersion demonstrates that all viscoelastic effects leading to the transition from the viscous to the elastic regime occur at Q values much lower than those covered by the *IXS* measurement in Ref. (Giordano and Monaco, 2010), which thus probe the substantially elastic or solid-like response of the sample. This evidence urged the authors to conclude that for the considered sample and thermodynamic conditions there's no need to assume the presence of further high frequency, microscopic relaxations as those extensively discussed in Chapter 8.

The one discussed above is an interpretative scheme complementary to viscoelastic approaches. Indeed, it describes the high-frequency spectrum of a liquid by generalizing the response of a crystal, and specifically, by introducing in it the structural disorder, while conversely, viscoelastic approaches generalize the hydrodynamic response of a liquid by including in it solid-like, or elastic, aspects.

As discussed, these two approaches are not mutually exclusive, as the "inherent disorder" interpretation described above mainly applies to the response of the liquid in the elastic limit, In this limit, the acoustic damping is dominated by static effects, like those stemming from the continuous distribution of structures, rather than the dynamic (ω-dependent) viscoelastic effects which dominate at longer times and distances.

9.6. Generalities on the propagation of a shear wave in a liquid

Let assume that a relatively weak periodic force is applied on a liquid. If the period of such force is much longer than the relaxation time, the liquid responds as a viscous system. In this regime, the fluid cannot support density wave propagation along a direction orthogonal to the applied force, and transverse density fluctuation generated by the force has a merely diffusive, or overdamped, nature. This overdamping owes to the extremely rapid internal rearrangements of adjacent layers of the liquid, which dissipate the effect of shear restoring forces.

If conversely, the period of the applied force becomes much shorter than the relaxation time (elastic limit), virtually no rearrangement can happen in the system within a single oscillation of the force; in this regime, solid-like aspects of the response emerge, and the liquid may support shear, or transverse propagation, albeit only for short distances.

It is natural to introduce an extinction length for the transverse mode, say d_e, which is defined as the longer distance travelled by the shear wave before being damped-off. It is to be anticipated that a scattering experiment on a liquid can best probe acoustic transverse modes when $Qd_e, \omega_t(Q)\tau(Q) \gg 1$, where $\omega_t(Q)$ is the transverse acoustic frequency. These conditions can be fulfilled, for instance, by changing either of the variables Q and T to access to the elastic or solid-like regime of the fluid response.

In principle, the circumstance that an amorphous system lacks privileged symmetry axes makes the polarization of an acoustic wave ill defined over scales comparable with first neighboring atoms' separations; therefore, symmetry arguments leading to a neat discrimination of the polarity of density waves lose physical ground at these scales. Nonetheless, it is still customary to attribute a longitudinal or transverse polarization to a given acoustic mode, if such a mode dominates the spectrum of velocity correlation with the corresponding polarization (see Chapter 2, par 2.8).

Notice that the ability of a fluid to support shear wave propagation in itself does not necessarily imply the onset of such a shear mode in the spectrum of density fluctuations. For instance, although noble gases and Lennard-Jones systems can support transverse acoustic propagation at low temperatures and sufficiently high Q's (Levesque and Verlet, 1970), no evidence of these modes was ever found found in the spectrum of density fluctuations. In fact, density fluctuations primarily couple with longitudinal movements only, and the appearance of a shear mode in the $S(Q, \omega)$ profile can only occur indirectly via some entanglement or mixing between modes of longitudinal and transverse polarization, customarily referred to as longitudinal-transverse coupling, or LTC (Sampoli *et al.*, 1997).

Expectedly, what hampers the emergence of distinct shear modes in the $S(Q, \omega)$ of a liquid is the weak amplitude of shear restoring forces due to both the lack of periodicity and, at sufficiently low frequencies, by dispersion effects induced by the system's rearrangements. Consequently, in fluid systems, transverse contributions to $S(Q, \omega)$, when observable at all, have substantial relative damping, thus typically appearing in the form of loosely resolved spectral features.

9.6.1. *A transverse mode in the spectrum of water*

The peculiar behavior of the terahertz dynamics of water attracted a considerable interest after the evidence of an additional low-frequency mode in the $S(Q, \omega)$, which was assigned to the onset of a shear mode propagation (Sette *et al.*, 1996); in this early *IXS* measurement it was reported that a transverse mode emerges in the spectrum of water only above $Q \approx 4 \, \text{nm}^{-1}$. The assignment of this spectral feature to a shear mode propagation was further substantiated by an *IXS* measurement jointly performed on water and ice along different crystallographic directions (Ruocco *et al.*, 1996). Also, based upon previous lattice vibrations calculations and *INS* measurements (Renker, 1969) the low frequency inelastic mode in the water spectrum was assigned to a transverse optical (TO) mode.

As a matter of fact, the evidence of an additional low-frequency peak in the water spectrum largely precedes the early *IXS* measurements by the group of Sette at ESRF, dating back to the seminal *MD* work of Rahman and Stillinger in the mid 1970s (Rahman and Stillinger, 1974) and the successive *INS* measurement in D_2O (Bosi *et al.*, 1978). Back then, this low-frequency feature was erroneously interpreted as the ordinary hydrodynamic sound mode, and assumed to coexist with a "fast sound" mode at higher frequency (Teixeira *et al.*, 1985). The respective assignment of the low- and the high-frequency modes in the water spectrum to a transverse and a longitudinal acoustic modes is one of the major scientific achievements of high-resolution *IXS* in its early development phase.

The coupling between density fluctuations and shear mode propagation in water was initially inferred by Balucani and collaborators (Balucani *et al.*, 1996), whose *MD* simulation showed that the longitudinal and transverse current spectra (see Eqs. 2.87 and 2.88) are dominated by the same two modes.

The successive *MD* simulation work by Sampoli and collaborators (Sampoli *et al.*, 1997) elucidated several important aspects associated to the *LTC* in water through a detailed study of the Q-evolution of current spectra. Specifically, this computational study demonstrated the occurrence of a polarization mixing of the two acoustic modes of water. It also revealed unambiguously that at sufficiently low Q's each of the two current spectra is dominated by a single distinct mode, which thus preserves, in this low Q-range, a better-defined polarization character. The simulation was carried out following the trajectories of 4000 D_2O molecules interacting through an extended simple point charge model, *SPC/E* (Berendsen, Grigera *et al.* 1987) potential simulating microscopic interactions in water at a temperature of 250 K, which corresponds to the maximum density for the used potential model. Few representative transverse and longitudinal current spectra computed by Sampoli and collaborators at various Q values are reported in Figure 9-10.

It appears therein that, at sufficiently large Q values, two broad spectral features emerge at approximately the same energies in the two spectral shapes, which is a clear indication of the mixed polarization of these acoustic modes. However, the two spectral features have different relative amplitude, and only the dominant one emerges in either of the current spectra profiles at all Q values, including the lowest ones.

One may argue that each of the two modes keeps a definite polarization only in the limited low Q-range where it appear as a

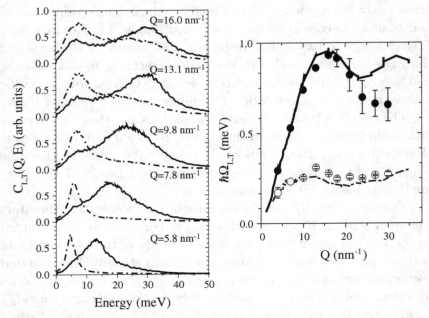

Figure 9-10: Left plot: MD simulation of the longitudinal (solid lines) and transverse (dot-dashed lines) current spectra are reported as computed by Sampoli and collaborators (Sampoli, Ruocco *et al.* 1997) at few typical *Q* values. Right plot: the corresponding dispersion curves of the longitudinal and transverse modes (same lines as in the left plot) are compared with the positions of the high (dots) and low frequency (open circles) peaks in the spectrum measured by *IXS* (Pontecorvo *et al.*, 2005).

single peak. However, Figure 9-10 also shows that longitudinal and transverse current spectra are consistently dominated by a high- and a low-frequency peak, respectively. With little ambiguity, we can thus assign to these peaks, a (prevalently) longitudinal and transverse character, respectively.

Also, the two acoustic modes exhibit clearly distinctive *Q*-dispersions that of the longitudinal mode being quite more pronounced at all *Q*'s.

Also, the two acoustic modes exhibit clearly distinctive trends; for instance, the dominant mode in the longitudinal currents has, as expected, systematically larger energy and a more pronounced *Q*-dependence.

The dispersion curves of the longitudinal and transverse modes defined as specified above are reported in the right plot of the same figure along with the ones determined from the analysis of the double-peaked structure of *IXS* spectra of water in Ref. (Pontecorvo, Krisch *et al.* 2005). It appears that simulated and measured curves are mutually consistent. This evidence supports the transverse nature of the low frequency mode in the *IXS* spectrum of water, thereby confirming the occurrence of an *LTC*.

The *IXS* work of Pontecorvo and collaborators covered broad thermodynamic and Q windows showing that the *LTC* could only occur when density fluctuations propagate at frequencies exceeding the inverse of the relaxation time, that is when the dynamic response is in its elastic, or solid-like, regime. Here the propagating acoustic perturbation "perceives" the surrounding liquid medium as frozen, or solid-like, like in a glass.

The virtual lack of viscous dissipation typical of this regime makes shear restoring forces more effective in promoting transverse acoustic propagation. As discussed, this regime can be met either by increasing Q or by decreasing the temperature. Figure 9-11 provides an example of the Q evolution of *IXS* spectra measured in liquid water at a low temperature (263 K), where the transverse mode clearly emerges up to become prevalent at the highest Q's. The plots show that the enhancement of the low-frequency transverse mode upon Q increase is accompanied by the gradual disappearance of the high-frequency longitudinal mode, which at best appears as a broad spectral shoulder from low to moderate Q's. Indeed, for Q larger than the position of first $S(Q)$ maximum (slightly larger than 20 nm^{-1} in water) the low-frequency transverse mode largely dominates the inelastic wings of the dynamic structure factor, becoming the only inelastic feature clearly observable in the spectral shape.

In the *IXS* work of Cimatoribus and collaborators (Cimatoribus, Saccani *et al.* 2010) a comparison was proposed between the terahertz dynamics of water in the stable liquid, solid and supercooled phases. In particular, in supercooling conditions, water exhibits a complex spectral shape, which contains both high and low-frequency inelastic modes somehow reminiscent of the sharper phonon peaks dominating

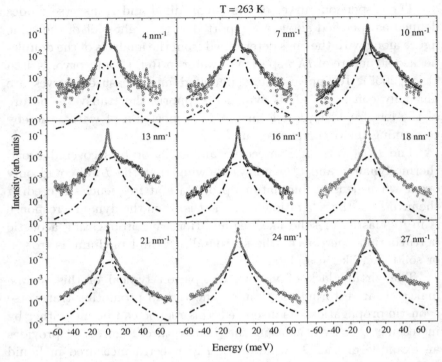

Figure 9-11: The Q-dependence of IXS spectra of water at $T = 263$ K. Raw data (circles) are compared with the best-fit lineshape (solid lines through data) and low-frequency mode (dot-dashed lines).

the spectrum of ice, also measured in the same work example of IXS spectra measured in these two sample phases are compared in Figure 9-12. Again, such a comparison stresses a vague similarity between the high-frequency response of a liquid and the one of its crystalline counterpart, also suggesting that, at low temperatures, the spectral shape is even more structured than reported by previous IXS measurements.

Figure 9-13 provides a representation of the link between a viscoelastic behavior and the onset of a shear mode propagation by displaying, in a double logarithmic plot, the sound dispersions of the longitudinal and the transverse mode computed in the MD work of Ref. (Sampoli, Ruocco et al. 1997) and already reported in Figure 9-10.

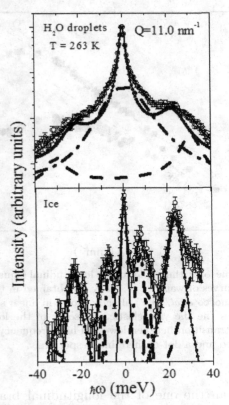

Figure 9-12:   IXS spectra measured in a droplet of supercooled water (upper panel) and in ice (lower panel). Data are from Ref. (Cimatoribus, Saccani *et al.* 2010) and are reported along with best-fit lineshapes and individual spectral modes.

The longitudinal branch follows a linear behavior with a slope coincident with the adiabatic sound velocity, as expected in the hydrodynamic limit. Upon increasing Q, the longitudinal branch undergoes the viscoelastic crossover, delimited by the dashed oval, characterized by a sharper, more-than-linear, Q-dependence.

This trend is the mentioned positive sound dispersion, revealing the coupling of density fluctuations with an active relaxation process. In the same Q-range, the second, transverse acoustic branch emerges (dashed lines). At its onset, the low-frequency branch seems to exhibit a linear Q-dependence, which at higher Q's becomes

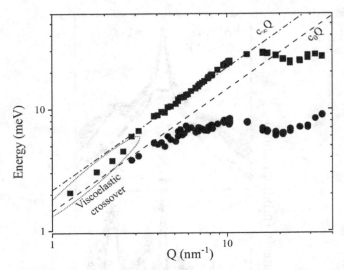

Figure 9-13:   Double logarithmic plot of the longitudinal (squares) and transverse (dots) dispersion curves of water at T 263 K, as obtained in the MD simulations of Ref. (Sampoli, Ruocco *et al.* 1997) already shown in linear scale in Figure 9-12. The ellipsis encloses the the viscoelastic, crossover of the longitudinal branch, which leads to the transition between low- and high-frequency linear trends, also reported as a dashed and a dot-dashed line, respectively.

much weaker than the one of the longitudinal branch. The latter trend manifests itself perhaps more clearly in the linear plot of Figure 9-10.

In conclusion, the plot seems to indicate that the emergence of the second peak happens at Q's at least as large as those characterizing the viscoelastic transition. This leaves still open the question on how, the lineshape evolves from the single to the double-shaped structure in the Q-range neighbouring the viscoelastic region. Owing to the small difference between the frequency of the longitudinal and the transverse modes in the viscoelastic region, addressing this question goes beyond the resolution capabilities of current *IXS* techniques. However, simultaneous *INS* and *IXS* investigations of the spectrum of deuterated water permitted to partially circumvent this limitation (Cunsolo *et al.*, 2012). Indeed, this joint measurement combined the superior resolution of *INS* spectrometers and the lack of kinematic limitations typical of *IXS* to achieve a precise determination of the spectral shape.

The use of Cold Neutron Chopper Spectrometer, *CNCS*, operating at Oak Ridge Laboratory permitted to measure the spectrum of D_2O with an extremely sharp resolution having a $.1meV$ FWHM and nearly Gaussian tails, which enabled the observation of the transverse mode even at ambient conditions. The joint use of an *IXS* spectrometer, Sector 30 of the Advanced Photon Source at Argonne National Laboratory, thanks to the virtual lack of kinematic limitation, permitted the parallel detection of the longitudinal mode emerging at higher frequencies in the spectrum of heavy water. One of the main results of this work is reported in Figure 9-14. The left plot displays the scattering signals $I'(Q,\omega)$ subtracted of the background and normalized to $S(Q)$, while the plot on the left compares the profiles $C'_L(Q,\omega) = (\omega/Q)^2 I'(Q,\omega)$, which can be considered as a normalized "experimental" analog of the current spectrum $C_L(Q,\omega) = (Q/\omega)^2 S(Q,\omega)$. The spectral contributions

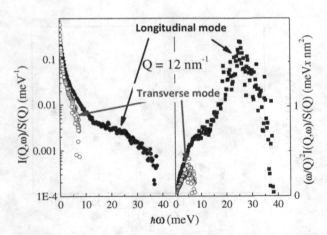

Figure 9-14: Right panel: A comparison between the ultra-high resolution ($<.1$ meV) INS and IXS measurement of the $S(Q,\omega)$ of D_2O at ambient conditions (open circles and black dots respectively). Left panel: the corresponding current spectra are reported for comparison. The dominant longitudinal and transverse peaks of the current spectra are indicated in the plot.

from longitudinal and transverse modes are also indicated by arrows in the plots.

The *INS* and *IXS* spectral shapes were analysed using a combined fitting procedure with a model based upon a memory function without thermal contribution, complemented by an additional *DHO* term (see Appendix 6A) accounting for the low-frequency transverse mode. This analysis led to the conclusion that, upon approaching the viscoelastic regime, the transverse mode becomes critically damped; also, the similar values of the undamped frequency and damping coefficient at low Q values were found comparable with the inverse of the relaxation time, as reported in Figure 9-15. In summary, in the viscoelastic regime, the shear mode yields an overdamped contribution to the spectrum not easily distinguishable from the relaxational, or Mountain, mode.

Although new *IXS* measurements are providing growing evidence of the same *LTC* phenomena in various liquid systems, as also discussed in the next section, several topics associated to it are still unsolved; they concern, for instance, the microscopic driving mechanism of this phenomenon. More in general, there is a critical

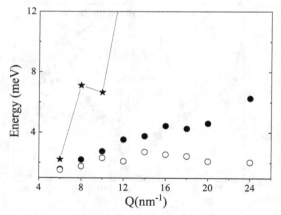

Figure 9-15:   The Q-dependence of the undamped frequency (dots) and the damping coefficient (circles) of the DHO model describing the transverse mode in the water spectrum are compared with the inverse of the relaxation time (line + stars). Data are from the joint *INS* and *IXS* experiment in Ref. (Cunsolo, Kodituwakku *et al.* 2012).

need for a firm unified theory accounting for the gradual emergence of a transverse dynamics in the spectrum of density fluctuations and explaining its link with the nature of microscopic interactions.

We conclude this brief excursus on literature *IXS* investigations of shear mode propagation in water by discussing the link between acoustic modes of liquid water and HB degrees of freedom, as originally inferred from Raman spectroscopy measurements in water (Walrafen, 1964; Walrafen, 1966). The longitudinal acoustic mode involves molecular center of mass translations hindered by the O-O stretching along the HB direction. Conversely, the transverse acoustic mode relates to centers of mass translations hindered by the bending of O-O-O triplets orthogonal to HBs. Interestingly, the activation energy derived from the temperature dependence of the stretching frequency is about 2.6 kcal/mole, a value comparable with the one derived from the Arrhenius plot of the structural relaxation time in Figure 8-14.

9.6.2. *The onset of a transverse dynamics in monatomic systems*

Besides the case of water, in the last few decades, the emergence of a transverse mode in the spectrum of density fluctuations is also being reported in a growing number of systems. These include SiO_2 (Dell'Anna *et al.*, 1998, Ruzicka *et al.*, 2004), GeO_2 (Bove *et al.*, 2005) and $GeSe_2$ (Orsingher *et al.*, 2010), all sharing with water a network of intermolecular bonds having a tetrahedral symmetry. This initially prompted the conjecture that the *LTC* could arise from the highly directional and intrinsically open character of the tetrahedral arrangement. However, it seems now well assessed that the *LTC* is not a prerogative of tetrahedral arrangements as it has been computationally observed also in an associated system whose network of bonds lacks a tetrahedral symmetry, such as glassy glycerol (Scopigno *et al.*, 2003).

Even more surprisingly, an analogous *LTC* was observed in monatomic and non-associated systems, as liquid metals. This is, for instance, the case of liquid Ga (Hosokawa *et al.*, 2008, Hosokawa *et al.*, 2009, Giordano and Monaco, 2011). As a matter of fact, this system is a rather peculiar liquid metal, due to the coexistence of metallic and covalent bonds in it, which gives rise to very short-living interatomic bonds, likely reminiscent of the crystalline form α-Ga (Gong *et al.*, 1993). In this perspective, it might not be surprising that a genuinely solid-like feature such as the onset of a transverse mode in the $S(Q, \omega)$ was reported by several *IXS* works on this sample.

In Ref. (Hosokawa *et al.*, 2009), the authors conjectured that the ability of liquid Ga to support transverse mode propagation could be partially due to solid-like cage effects acting at nanometer scales as a restoring force promoting the propagation of shear density waves. This ability is likely connected to the remarkably high value of the Poisson ratio of Ga, which turns out to be comparable to that of rubber-like materials.

An *IXS* work on sodium documented the presence in the spectrum of both longitudinal and transverse acoustic excitations at frequencies similar to their counterparts in the crystal (Giordano and Monaco 2010). In this work, already discussed in detail in Section 9.5.1, the disorder contribution to the acoustic damping in a liquid was represented assuming the existence of a continuous distribution of local structures. One might anticipate that, at low enough temperatures, such a distribution becomes so sharp — and the disorder-related transverse damping so small — that transverse fluctuations transform to underdamped propagating modes.

A study of transverse mode in Sn (Hosokawa *et al.*, 2013) suggested that inelastic modes in this metal are strongly localized being their propagation length smaller than the Ioffe-Riegel distance (Thouless, 1974, Taraskin and Elliott, 1997), which defines the maximum length scale for which phonons can still be considered well-defined excitations, *i.e.* the phonon mean free path $l = c_s(Q)/z_s(Q)$.

A more recent *INS* study on liquid Zn (Zanatta *et al.*, 2015) documented the presence of a low-frequency excitation with a transverse acoustic-like nature, which was connected to the peculiar anisotropic interactions of this system. This conclusion seems consistent with the anomalously high value estimated for the Poisson ratio of Zn, which is comparable to that of rubber-like materials.

Furthermore, a previously cited *IXS* work on sodium showed the presence in the spectrum of both longitudinal and transverse acoustic excitations at frequencies approaching to the corresponding ones in the crystal phase (Giordano and Monaco, 2011).

Although documented by many independent works, the presence of a transverse mode in the spectrum of liquid systems often manifests itself through barely resolved shoulders on the side of a dominating central peak. This partially owes to the broad resolution function of current *IXS* instruments, which often makes the support of parallel *MD* simulation necessary to convey unambiguous evidence of such a spectral mode. Figure 9-16 provides an example of how

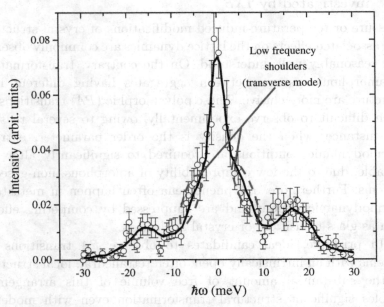

Figure 9-16:   The IXS spectrum of liquid Ga measured in Ref. (Hosokawa, Inui *et al.* 2011) (open circles). The raw measurement (open circles) is reported against the best-fitting lineshape (solid line) obtained with a model composed of two DHO profiles (see Appendix 6A) accounting for inelastic collective modes, plus a Lorentzian term accounting for the quasi-elastic central peak.

the spectral contribution from a transverse mode can appear in the *IXS* spectrum under the most favourable conditions. There, the *IXS* spectrum in Ga at 100 °C measured at $Q = 10$ nm^{-1} by Hosokawa and collaborators (Hosokawa *et al.*, 2008) is reported along with the best-fitting model lineshape including two *DHO* profiles (see Appendix 6A) and an additional quasi-elastic Lorentzian term.

The shear mode contribution to the spectrum manifests itself through the presence of two low-frequency shoulders on the side of the central peak. It is worth noticing that molten metals have a relatively weak central peak which provides a decisive advantage when focusing on low-frequency modes, manifests itself through the presence often obscured by the large quasielastic wings.

9.7. Polyamorphism phenomena in simple systems investigated by *IXS*

Pressure or temperature-induced modifications of crystal structures and associated effects on the lattice dynamics are commonly observed and reasonably well understood. On the contrary, transformations of amorphous systems between aggregates having different local structure are more elusive. These polyamorphic (*PA*) transitions are often difficult to observe experimentally, owing to several reasons. For instance, when the density is the order parameter, extreme thermodynamic conditions are required to significantly alter this variable, due to the low compressibility of amorphous, non-gaseous, systems. Furthermore, *PA* phenomena often happen in metastable thermodynamic regions and are suppressed by competing effects, such as glass transition, or crystal nucleation.

In principle, ideal candidates to observe *PA* transitions are systems with an intrinsically open, often tetrahedral, local structure, as the substantial amount of free volume of this arrangement enables significant structural transformation even with moderate thermodynamic changes (Brazhkin and Lyapin, 2003).

This circumstance explains the observation of *PA* effects in water (Mishima and Stanley, 1998), liquid silicon (Beye *et al.*, 2010), germanium (Koga *et al.*, 2002), and phosphorus (Katayama *et al.*,

2000), as well as in amorphous SiO_2 (Lacks, 2000) and GeO_2 (Itie *et al.*, 1989) . Despite thorough experimental scrutiny, some general aspects of *PA* transitions are still obscure. This includes, for instance, the possible influence of *PA* transformations on the propagation of collective modes of $S(Q, \omega)$.

Effects on the collective dynamics are supposedly more relevant at mesoscopic ($\approx$ nm) length scales, where dynamic events become strongly coupled with local atomic arrangements.

For instance, dynamic signatures of a *PA* transition were found in the *IXS* spectrum of liquid sulphur (Monaco *et al.*, 2005) and vitreous GeO_2 (Cunsolo *et al.*, 2015). In these works, changes in the dynamic response were sought for in the temperature- and the pressure dependence of the sound velocity, respectively, as derived from the slope of the low Q, essentially linear, dispersion. As readily emerges from Figure 9-18, the main effect of a *PA* transition in these materials is a slope discontinuity or cusp-like behavior in the T- and the P-dependence of sound velocity (left and right plots, respectively). This trend is emphasized by the comparison of the straight lines best fitting the pressure increase (and the temperature decrease) of the sound velocity before and after the slope discontinuity.

The behavior emerging from Figure 9-17 might appear surprising if one considers that the sound velocity reported therein is derived as the slope of the low Q dispersion. If viscoelastic effects occur at Q's lower than those covered by the measurement, such a velocity coincides with the elastic sound speed; at variance of the adiabatic sound the latter is not directly amenable to thermodynamic (macroscopic) properties of the system and its link with a *PA* transition is, in principle, more elusive. Along the same line, an earlier *IXS* measurement of the density dependence of the infinite frequency sound velocity of water (Krisch *et al.*, 2002), showed an anomalous kink at a density comparable to the one of the hypothesized second critical point (Mishima and Stanley, 1998).

In summary, the direct influence of thermodynamic anomalies on the short-time (high-frequency) transport properties of a fluid is a result calling for further interpretative efforts.

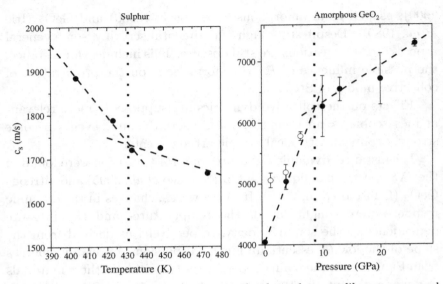

Figure 9-17:   Panels A) and B) respectively show the cusp-like pressure and temperature dependences of sound velocity (dots) measured by *IXS* methods on sulphur (Monaco *et al.*, 2005) and GeO$_2$ (Cunsolo *et al.*, 2015). Open circles represent data from Ref. (Bencivenga and Antonangeli, 2014), while dashed lines represent linear best fits of data before and after the *PA* crossover, whose position is indicated by a vertical dotted line.

In a more recent joint *ab initio MD* simulation and *IXS* measurement on Rb (Bryk *et al.*, 2013) a similar trend of the sound velocity was also reported and ascribed to the occurrence of a liquid-liquid transition. The *IXS* measurements were executed at up to 6.6 GPa at 573 K, and this pressure range was successively extended to 27.4 GPa by ab initio molecular dynamics simulations. The analysis of the P-dependence of the simulated $g(r)$ revealed a more-than-linear reduction of the atomic nearest neighboring distance above 12.4 GPa. The pressure dependence of sound velocity is reported in Figure 9-18 as derived from the slope of the lowest Q dispersion of the main peak of simulated $C_L(Q, \omega)$ profiles or from their *IXS* experimental analogous. Figure 9-18 shows that this trend is also accompanied by an anomalous pressure dependence of the Rb atom diffusivity, which at pressure between 10 GPa and 13 GPa, changes slope. This trend indicates a substantial reduction of the core radius of Rb atoms, as

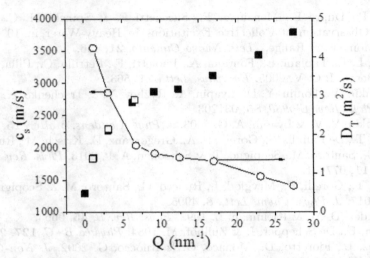

Figure 9-18: The sound velocity (full squares) and the diffusion coefficient (lines + circles, referring to the y-axis on the right) determined in an *ab initio MD* simulation on a model representative of liquid Rb along the $T = 573$ K isothermal path; *IXS* measurements on the same system are also reported as open squares. The plot is redrawn from Ref. (Bryk, De Panfilis *et al.* 2013).

it follows from a relation between the particle size and diffusivity (Bylander and Kleinman, 1992).

References

Angelini, R., Giura, P., Monaco, G., Ruocco, G., Sette, F. & Verbeni, R. 2002. *Phys. Rev. Lett.*, **88**, 255503.

Balucani, U. & Zoppi, M. 1994. *Dynamics of the Liquid State*, Oxford, Clarendon Press.

Bencivenga, F. & Antonangeli, D. 2014. *Phys. Rev. B*, **90**, 134310.

Bencivenga, F., Cunsolo, A., Krisch, M., Monaco, G., Orsingher, L., Ruocco, G., Sette, F. & Vispa, A. 2007. *Phys. Rev. Lett.*, **98**, 085501.

Beye, M., Sorgenfrei, F., Schlotter, W. F., Wurth, W. & Fohlisch, A. 2010. *Proc. Natl. Acad. Sci. U.S.A.*, **107**, 16772.

Bolmatov, D., Brazhkin, V. V. & Trachenko, K. 2013. *Nat. Commun.*, **4**, 2331.

Bolmatov, D., Zhernenkov, M., Zav'yalov, D., Stoupin, S., Cai, Y. Q. & Cunsolo, A. 2015a. *J. Phys. Chem. Lett.*, **6**, 3048.

Bolmatov, D., Zhernenkov, M., Zav'yalov, D., Tkachev, S. N., Cunsolo, A. & Cai, Y. Q. 2015b. *Sci. Rep.*, **5**, 15850.

Bosi P., Dupre F., Menzinger F., Sacchetti F. & Spinelli M. C. 1978. "Observation of Collective Excitations in Heavy-Water in 10^8 cm^{-1} Momentum Range," *Lett. Nuovo Cimento*, **21**, 436.

Bove, L. E., Fabiani, E., Fontana, A., Paoletti, F., Petrillo, C., Pilla, O. & Bento, I. C. V. 2005. *Europhys. Lett.*, **71**, 563.

Brazhkin, V., Fomin, Y. D., Lyapin, A., Ryzhov, V. & Trachenko, K. 2012. *Phys. Rev. E y980*, **85**, 031203.

Brazhkin, V. V. & Lyapin, A. G. 2003. *J. Phys. Condens. Matter*, **15**, 6059.

Bryk, T., De Panfilis, S., Gorelli, F. A., Gregoryanz, E., Krisch, M., Ruocco, G., Santoro, M., Scopigno, T. & Seitsonen, A. P. 2013. *Phys. Rev. Lett.*, **111**, 077801.

Bryk, T., Gorelli, F., Mryglod, I., Ruocco, G., Santoro, M. & Scopigno, T. 2017. *J. Phys. Chem. Lett.*, **8**, 4995.

Bylander, D. M. & Kleinman, L. 1992. *Phys. Rev. B*, **45**, 9663.

Cohen, E., De Schepper, I. & Zuilhof, M. 1984. *Physica B+C*, **127**, 282.

Comez, L., Fioretto, D., Monaco, G. & Ruocco, G. 2002. *J. Non-Cryst. Solids*, **307–310**, 148.

Crevecoeur, R. M., Verberg, R., de Schepper, I. M., de Graaf, L. A. & Montfrooij, W. 1995. *Phys. Rev. Lett.*, **74**, 5052.

Cunsolo, A. 1999. Thesis, Universite' J. Fourier.

Cunsolo, A., Li, Y., Kodituwakku, C. N., Wang, S., Antonangeli, D., Bencivenga, F., Battistoni, A., Verbeni, R., Tsutsui, S., Baron, A. Q., Mao, H. K., Bolmatov, D. & Cai, Y. Q. 2015. *Sci. 1984a Rep.*, **5**, 14996.

Cunsolo, A., Pratesi, G., Ruocco, G., Sampoli, M., Sette, F., Verbeni, R., Barocchi, F., Krisch, M., Masciovecchio, C. & Nardone, M. 1998. *Phys. Rev. Lett.*, **80**, 3515.

Cunsolo, A., Pratesi, G., Verbeni, R., Colognesi, D., Masciovecchio, C., Monaco, G., Ruocco, G. & Sette, F. 2001. *J. Chem. Phys.*, **114**, 2259.

Cunsolo, A., Ruocco, G., Sette, F., Masciovecchio, C., Mermet, A., Monaco, G., Sampoli, M. & Verbeni, R. 1999. *Phys. Rev. Lett.*, **82**, 775.

de Schepper, I., Verkerk, P., van Well, A., de Graaf, L. & Cohen, E. *Phys. Rev. Lett.*, **53**, 402.

de Schepper, I. M., Cohen, E. G. D., Bruin, C., Vanrijs, J. C., Montfrooij, W. & Degraaf, L. A. 1988. *Phys. Rev. A*, **38**, 271.

de Schepper, I. M., van Rijs, J. C., van Well, A. A., Verkerk, P., de Graaf, L. A. & Bruin, C. 1984b. *Phys. Rev. A*, **29**, 1602.

de Schepper, I. M., Verkerk, P., van Well, A. A. & de Graaf, L. A. 1984c. *Phys. Lett. A*, **104**, 29.

Dell'Anna, R., Ruocco, G., Sampoli, M. & Viliani, G. 1998. *Phys. Rev. Lett.*, **80**, 1236.

Fioretto, D., Buchenau, U., Comez, L., Sokolov, A., Masciovecchio, C., Mermet, A., Ruocco, G., Sette, F., Willner, L., Frick, B., Richter, D.

& Verdini, L. 1999. *Physical Review E, Statistical, Nonlinear, and Soft Matter Physics*, **59**, 4470.

Frenkel, J. 1947. *Kinetic Theory of Liquids*, Oxford University Press, Oxford.

Giordano, V. M. & Monaco, G. 2010. *Proc. Natl. Acad. Sci. U.S.A.*, **107**, 21985.

Giordano, V. M. & Monaco, G. 2011. *Phys. Rev. B*, **84**.

Gong, X. G., Chiarotti, G. L., Parrinello, M. & Tosatti, E. 1993. *Europhys. Lett.*, **21**, 469.

Gorelli, F., Santoro, M., Scopigno, T., Krisch, M. & Ruocco, G. 2006. *Phys. Rev. Lett.*, **97**, 245702.

Gotze, W. 2008. *Complex Dynamics of Glass-Forming Liquids: A Mode-Coupling Theory*, OUP, Oxford.

Gotze, W. & Mayr, M. R. 2000. *Phys. Rev. E*, **61**, 587.

Hosokawa, S., Inui, M., Kajihara, Y., Matsuda, K., Ichitsubo, T., Pilgrim, W. C., Sinn, H., Gonzalez, L. E., Gonzalez, D. J., Tsutsui, S. & Baron, A. Q. 2009. *Phys. Rev. Lett.*, **102**, 105502.

Hosokawa, S., Munejiri, S., Inui, M., Kajihara, Y., Pilgrim, W. C., Ohmasa, Y., Tsutsui, S., Baron, A. Q., Shimojo, F. & Hoshino, K. 2013. *J. Phys. Condens. Matter*, **25**, 112101.

Hosokawa, S., Pilgrim, W. C., Sinn, H. & Alp, E. E. 2008. *J. Phys. Condens. Matter*, **20**, 114107.

Itie, J., Polian, A., Calas, G., Petiau, J., Fontaine, A. & Tolentino, H. 1989. *Phys. Rev. Lett.*, **63**, 398.

Jayaraman, A. 1986. Ultrahigh pressures. *Review of Scientific Instruments*, **57**(6): 1013–1031.

Katayama, Y., Mizutani, T., Utsumi, W., Shimomura, O., Yamakata, M. & Funakoshi, K. 2000. *Nature*, **403**, 170.

Koga, J., Okumura, H., Nishio, K., Yamaguchi, T. & Yonezawa, F. 2002. *Phys. Rev. B*, **66**.

Krisch, M., Loubeyre, P., Ruocco, G., Sette, F., Cunsolo, A., D'Astuto, M., LeToullec, R., Lorenzen, M., Mermet, A., Monaco, G. & Verbeni, R. 2002. *Phys. Rev. Lett.*, **89**, 125502

Lacks, D. J. 2000. *Phys. Rev. Lett.*, **84**, 4629.

Levesque, D. & Verlet, L. 1970. *Phys. Rev. A*, **2**, 2514.

Lewis, J. W. E. & Lovesey S. W. 1977. *J. Phys. C*, **10** 3221.

Lovesey, S. 1984. *Phys. Rev. Lett.*, **53**, 401.

McGreevy, R. L. & Mitchell, E. W. J. 1985. *J. Phys. C Solid State Phys.*, **18**, 1163.

Mishima, O. & Stanley, H. E. 1998. *Nature*, **396**, 329.

Monaco, G., Crapanzano, L., Bellissent, R., Crichton, W., Fioretto, D., Mezouar, M., Scarponi, F. & Verbeni, R. 2005. *Phys. Rev. Lett.*, **95**, 255502.

Monaco, G., Cunsolo, A., Ruocco, G. & Sette, F. 1999. *Phys. Rev. E*, **60**, 5505.

Mishima, O. & Stanley, H. 1998. The relationship between liquid, super-cooled and glassy water. *Nature*, **396**, 329.

Montfrooij, W., de Graaf, L. A. & de Schepper, I. M. 1992. *Phys. Rev. B*, **45**, 3111.

NIST Thermodynamic data are takn from the NIST database: http:// webbook.nist.gov/chemistry/form-ser.html.

Orsingher, L., Baldi, G., Fontana, A., Bove, L., Unruh, T., Orecchini, A., Petrillo, C., Violini, N. & Sacchetti, F. 2010. *Phys. Rev. B*, **82**, 115201.

Renker, B. 1969. "Phonon Dispersion in D_2O-Ice," *Phys. Lett. A*, **30**, 493.

Ruocco, G., Sette, F., Bergmann, U., Krisch, M., Masciovecchio, C., Mazzacurati, V., Signorelli, V. & Verbeni, R. 1996. "Equivalence of the sound velocity in water and ice at mesoscopic wavelengths," *Nature*, **379**, 521–523 (1996).

Ruzicka, B., Scopigno, T., Caponi, S., Fontana, A., Pilla, O., Giura, P., Monaco, G., Pontecorvo, E., Ruocco, G. & Sette, F. 2004. *Phys. Rev. B*, **69**.

Sampoli, M., Bafile, U., Barocchi, F., Guarini, E. & Venturi, G. 2008. *J. Phys. Condens. Matter*, **20**, 104206.

Scarponi, F., Comez, L., Fioretto, D. & Palmieri, L. 2004. *Phys. Rev. B*, **70**.

Scopigno, T., D'Astuto, M., Krisch, M., Ruocco, G. & Sette, F. 2001. *Phys. Rev. B*, **64**, 012301.

Scopigno, T., Pontecorvo, E., Di Leonardo, R., Krisch, M., Monaco, G., Ruocco, G., Ruzicka, B. & Sette, F. 2003. *J. Phys. Condens. Matter*, **15**, S1269.

Scopigno, T., Ruocco, G. & Sette, F. 2005. *Rev. Mod. Phys.*, **77**, 881.

Simeoni, G. G., Bryk, T., Gorelli, F. A., Krisch, M., Ruocco, G., Santoro, M. & Scopigno, T. 2010. *Nat. Phys.*, **6**, 503.

Taraskin, S. N. & Elliott, S. R. 1997. *Europhys. Lett.*, **39**, 37.

Teixeira J., Bellissent-Funel M. C., Chen S. H. & Dorner B. 1985. "Observation of new short-wavelength collective excitations in heavy water by coherent inelastic neutron scattering," *Phys. Rev. Lett.*, **54**, 2681.

Thouless, D. J. 1974. *Phys. Rep.*, **13**, 93.

van Well, A. A. & de Graaf, L. A. 1985. *Phys. Rev. A*, **32**, 2396.

Verbeni, R., Cunsolo, A., Pratesi, G., Monaco, G., Rosica, F., Masciovecchio, C., Nardone, M., Ruocco, G., Sette, F. & Albergamo, F. 2001. *Phys. Rev. E*, **64**, 021203.

Verkerk, P., van Well, A. A. & de Schepper, I. M. 1987. *J. Phys. C Solid State Phys.*, **20**, L979.

Walrafen, G. E. 1964. "Raman Spectral Studies of Water Structure." *J. Chem Phys.*, **40**, 3249.

Walrafen, G. E. 1966. "Raman Spectral Studies of the Effects of Temperature on Water and Electrolyte Solutions." *J. Chem Phys.*, **44** 1546.

Zanatta, M., Sacchetti, F., Guarini, E., Orecchini, A., Paciaroni, A., Sani, L. & Petrillo, C. 2015. *Phys. Rev. Lett.*, **114**, 187801.

Conclusive remarks

In summary, this book aimed at discussing both practical and fundamental aspects of inelastic X-Ray scattering (*IXS*), also providing examples of how this spectroscopic method can be successfully applied to investigate the terahertz dynamics of simple fluids. Although the focus was on monatomic systems, we also discussed the case of water, due to its crucial role in *IXS* studies of structural relaxations. Furthermore, the isotropic distribution of the core electronic cloud in H_2O molecules minimizes the spectral contributions from molecular rotations, thus making the *IXS* response of water directly comparable to that of monatomic systems. Conversely, we discarded the case of systems of molecules with anisotropic core electron charge distribution, as it poses significant theoretical challenges, partly eluding the scope of the book. Also, glasses and glass-forming materials were not covered by this book, despite the enormous interest they attracted inside the *IXS* community. Indeed, the spectral behavior of these materials would be, in itself, worth a dedicated monograph, possibly providing a useful tutorial for new generations of scientists picking up this field of research.

Hopefully, the unavoidably incomplete selection of experimental topics discussed in this book provided the reader with a flavor of some significant advances brought about by still young technique in our knowledge of the terahertz dynamic response of simple systems. These include the opportunity of capturing, with a single spectrometer, the evolution of the spectral shape through the whole dynamic crossover between hydrodynamic and single-particle regimes while also shedding insight on the gradual onset of quantum effects. The improved count rates and lack of kinematic limitations disclosed a whole new class of inelastic scattering investigations affording detailed and physically informative modeling of the lineshape. Many of these studies proposed a lineshape description derived within the memory function formalism framework, shedding new light onto relaxation processes in liquid and supercritical systems and their effect on the spectrum of density fluctuations. More recently, substantial technical advances have enabled *IXS* measurements in extreme thermodynamic regions, whose outcome calls for a global revision of the broadly agreed notion of the supercritical phase as an inherently uniform condition of matter, suggesting instead the existence of distinct thermodynamic domains in which the fluid's response resembles either the one of a viscoelastic liquid or that of a compressed gas. Most recent *IXS* investigations of the high-frequency dynamics of fluids in the elastic or solid-like short-time regime demonstrated the possibility of identifying the various contributions to the acoustic damping in a liquid. Among various solid-like features exhibited by the high-frequency dynamics of fluids, one of the most frequently reported is the emergence of a transverse dynamics in the spectrum of density fluctuations. Finally, an increasing number of *IXS* results showed the potentialities of this technique in elucidating dynamical aspects associated with liquid-liquid and glass-glass poly(a)morphic transitions. Overall, many of these results have changed our commonsense perception of the liquid phase by emphasizing some similarities between the high-frequency behavior of viscoelastic fluids and that of amorphous solids.

Looking ahead, it would be useful to discuss here possible future trends in *IXS* research. Rather than providing a comprehensive list, I would focus on a single topic which, in my opinion, holds the promise of bringing new excitement to the field. I will also discuss technical and interpretative challenges to be faced while investigating this topic.

Using IXS as a tool to advance terahertz phononics

Across the years, the mainstream interest of the *IXS* community has slowly, yet systematically, drifted from simple monatomic systems to more complex ones as associated or glass forming materials. In the last years an increasing attention was devoted to complex Soft Matter system presenting a nanoscale organization. In my opinion, artificial structures as nanoparticles (*NPs*) superlattices are of particular interest due to the high flexibility offered by their structural design, which lends itself to the discovery of new functionalities. Among them, especially appealing appears the possibility of controlling and manipulating sound propagation, which is an emerging research field, usually referred to as phononics. The extension of this field to the mostly uncharted terahertz territory is of particular interest, as terahertz phonons are the leading carriers of heat transport in dielectrics, and their shaping is essential to the development of a whole class of novel thermal devices (Maldovan 2013). Terahertz phononics can be implemented by designing nanometer-spaced *NP* superlattices to function as phononic crystals (*PCs*), *i.e.* structures enabling to steer acoustic propagation, possibly stopping it in specific frequency bands called "phononic gaps" (Liu, Zhang *et al.* 2000, Schneider, Liaqat *et al.* 2012). Most importantly, the properties of PCs are crucial for the development of thermal diodes (Li, Wang *et al.* 2004), energy-harvesting systems (Lv, Tian *et al.* 2013), and other thermal apparatuses (Liu, Zhang *et al.* 2000, Hopkins, Reinke *et al.* 2011).

Furthermore, executing *IXS* measurements on *NP* superlattices in a broad Q-range offers the opportunity of probing dynamic events occurring over different scales. Specifically, when Q's approaches the inverse of nearest neighbor NP separations, one can probe interparticle phonons involving superlattice vibrations in an ordered *NP* assembly. For Q's at least as large as the inverse of the single *NP* size, phonons propagating through the *NP* interior can also emerge in the spectrum.

Overall, a fascinating aspect of this emerging field of research is the reconsideration of the primary scientist duty from the mere observation of physical phenomena, to the actual control over these phenomena through the project, design and creation of artificial structures. In *NP* superlattices, everything is in principle tunable, including relevant *NP* features — like mass, size, and shape- and bonding parameters — such as linkage strength and distance. For instance, connections between *NPs* can consist of *DNA* strands whose length and stiffness can be easily tuned (Nykypanchuk, Maye *et al.* 2008).

Unfortunately, this field is still in its infancy, only a few partial and very preliminary *IXS* results having been reported to date. Among them, it is worth mentioning a recent experiment on diluted aqueous Au-*NP* suspensions demonstrating that even a very sparse concentration of *NPs* in immersion can damp off acoustic-like excitations propagating in pure liquid water (De Francesco, Scaccia *et al.* 2018). A recent *IXS* work suggested the possibility of shaping the acoustic response of liquid water upon confinement in aligned carbon nanotubes. Specifically, it was observed that the quasi-unidimensional geometry selects the axial propagation of transverse acoustic modes of water.

The critical role of the spectrometer performance

Regardless of the specific topic pursued by the experiment, *IXS* studies on nanostructured systems might soon encounter an *impasse*, due to the limited coverage of the low Q, ω region where the spectral contributions from collective modes are best observable. The access to this *"no man's land"* imposes strict demands on both energy and momentum resolutions of the measurement. Substantial advances in this area would require the use of next-generation *IXS* spectrometers implementing new monochromatization/energy analysis concepts, like those described in Chapter 4 (Shvyd'ko, Lerche *et al.* 2006). These instruments can guarantee the superior performance required to accurately discern the low-frequency portion of the spectral shape, directly informative of the slow collective dynamics of nanostructured systems. Irrespective of the availability of these novel spectrometers, there is still much one can do to achieve a more robust interpretation of the experimental outcome.

Revisiting the lineshape analysis

As mentioned, lineshape analyses of *IXS* measurements typically rest on phenomenological modeling of measured spectral profiles, in which the Q-dependence of shape parameters is determined *a posteriori* only by best-fitting spectra measured at various Q's. In principle, the outcome of this approach may critically depend on the model adopted. Also, mesoscale heterogeneities in nanostructured systems add further complexity, possibly originating additional, loosely resolved, low-frequency modes, and causing a hardly predictable Q-dependence of shape parameters. In this fuzzy scenario, it becomes essential to minimize the inherent bias of the modeling, ideally opting for inferential methods which: (1) establish how competitive models perform in describing the measurement, (2) inherently favour models with fewer adjustable parameters, (3) avoid convergence to local, instead of global, chi-square minima, (4) make use of minimally invasive external constraints and, most importantly, (5) efficiently cope with the limited statistical accuracy of experimental results.

Bayesian inferential methods (Sivia and Skilling 2006, Gelman, Carlin *et al.* 2013) can fulfill all these requirements. When applied to the modeling of spectroscopic results, these methods can enable probabilistic hypothesis tests involving, for instance, the number of inelastic modes most likely to appear in the spectrum. This unknown number, or, possibly, the analytical model function can be optimized to best describe measurements at hand (De Francesco, Guarini *et al.* 2016).

In general, the use of a Bayesian inference analysis provides valuable help whenever dealing with spectral signal noisy or characterized by spectral features highly damped, weak, or ill-resolved by the measurement. Bayesian inference offers a robust method to make sensible predictions based on an individual experimental outcome. This probabilistic inference provides robust support to the interpretation of evidence acquired in unique events, as, *e.g.*, a single spectral acquisition. Also, the Bayesian analysis estimates model parameters through averages rather than optimization procedures, its outcome thus being naturally protected against the deceiving proximity of fictitious (local, rather than global) optima. Most importantly, unnecessarily complicated models are penalized by the assignment of low weight (likelyhood) in the average. Indeed, Bayesian inference naturally incorporates the Occam's razor principle in the method (MacKay, 2003). According to this principle, among competitive hypotheses providing a satisfactory account of experimental evidence, the simplest one is to be privileged. Of course, the ultimate scope of this approach is not surrogating the investigator judgment, to whom is ultimately entrusted any decision on the significance of the inference outcome. In general, the use of a Bayesian inference analysis provides valuable help whenever dealing with spectral signal noisy or characterized by spectral features highly damped, weak, or ill-resolved by the measurement. Bayesian inference offers a robust method to make sensible predictions based on evidence acquired in unique events, as, *e.g.*, a single spectral acquisition.

The critical role of molecular dynamic simulations

A typical Soft Matter system can contain several million atoms and undergo long-lasting phase transitions, thus being hardly simulated by atomistic Molecular Dynamics (*MD*) simulations. However, advanced computers with petaflop capabilities can afford coarse-grained simulations of these systems, thus possibly providing strong interpretative support to spectroscopic measurements. Another valuable help from computer simulations is the possibility to perform "virtual experiments" like those recently developed for neutron scattering (Udby, Willendrup *et al.* 2011). These new codes combine ray-tracing methods — mimicking the effect of instrument optics — and MD simulations — accounting for the sample response. Among various advantages, they would enable a better assessment of the feasibility of a planned experiment, thus possibly making more efficient the use of allocated beamtime in synchrotron research facilities.

References

De Francesco, A., E. Guarini, U. Bafile, F. Formisano and L. Scaccia (2016). "Bayesian approach to the analysis of neutron Brillouin scattering data on liquid metals." *Phys. Rev. E* **94**: 023305.

De Francesco, A., L. Scaccia, M. Maccarini, F. Formisano, Y. Zhang, O. Gang, D. Nykypanchuk, A. H. Said, B. M. Leu and A. Alatas "Damping Off Terahertz Sound Modes of a Liquid upon Immersion of Nanoparticles." *ACS Nano* **12**, 8867 (2018).

De Francesco, A., L. Scaccia, F. Formisano, M. Maccarini, F. De Luca, A. Parmentier, A. Alatas, A. Suvorov, Y. Q. Cai and R. Li "Shaping the terahertz sound propagation in water under highly directional confinement." *Phys. Rev. B* **101**, 054306. (2020).

lman, A., J. B. Carlin, H. S. Stern, D. B. Dunson, A. Vehtari and D. B. Rubin (2013). *Bayesian data analysis.* Boca Raton, USA, CRC press.

Hopkins, P. E., C. M. Reinke, M. F. Su, R. H. Olsson, E. A. Shaner, Z. C. Leseman, J. R. Serrano, L. M. Phinney and I. El-Kady "Reduction in the thermal conductivity of single crystalline silicon by phononic crystal patterning." *Nano Lett.* **11**(1): 107 (2011).

Li, B., L. Wang and G. Casati. "Thermal diode: Rectification of heat flux." *Phys. Rev. Lett.* **93**, 184301 (2004).

Liu, Z., X. Zhang, Y. Mao, Y. Zhu, Z. Yang, C. T. Chan and P. Sheng "Locally resonant sonic materials." *Science* **289** 1734 (2000).

Lv, H., X. Tian, M. Y. Wang and D. Li "Vibration energy harvesting using a phononic crystal with point defect states." *Appl. Phys. Lett.* **102**, 034103 (2013).

MacKay D. J. Information theory, inference and learning algorithms, Cambridge University Press, Cambridge (2003).

Maldovan, M. "Sound and heat revolutions in phononics." *Nature* **503** 209 (2013).

Nykypanchuk, D., M. M. Maye, D. Van Der Lelie and O. Gang "DNA-guided crystallization of colloidal nanoparticles." *Nature* **451**, 549 (2008).

Schneider, D., F. Liaqat, E. H. El Boudouti, Y. El Hassouani, B. Djafari-Rouhani, W. Tremel, H.-J. R. Butt and G. Fytas "Engineering the hypersonic phononic band gap of hybrid Bragg stacks." *Nano Lett.* **12**, 3101 (2012).

Shvyd'ko, Y. V., M. Lerche, U. Kuetgens, H. D. Ruter, A. Alatas and J. Zhao "X-ray Bragg diffraction in asymmetric backscattering geometry." *Phys. Rev. Lett.* **97** 235502 (2006).

Sivia, D. and J. Skilling *"Data analysis: a Bayesian tutorial."* Oxford, Oxford University Press (2006).

Udby, L., P. K. Willendrup, E. Knudsen, C. Niedermayer, U. Filges, N. B. Christensen, E. Farhi, B. Wells and K. Lefmann "Analysing neutron scattering data using McStas virtual experiments." *Nucl. Inst. Meth. A* **634**, S138 (2011).

Printed in the United States
by Baker & Taylor Publisher Services